D1269115

CINCINNATI STATE
WITHDRAWN

A MANUAL OF GEOLOGY FOR CIVIL ENGINEERS

A Manual of
Geology for Civil Engineers

JOHN PITTS
Nanyang Technological Institute, Singapore

A HALSTED PRESS BOOK

John Wiley & Sons
New York–Toronto

624.15
P692

Published in the U.S.A., Canada
and Latin America by Halsted Press,
a Division of John Wiley & Sons, Inc.,
New York.

Published in Singapore by
World Scientific Publishing Co. Pte. Ltd., 1984

Library of Congress Cataloging in Publication Data

Pitts, John.
A manual of geology for civil engineers.

"A Halsted Press book."
1. Engineering geology. I. Title.
TA705.P49 1984 624.1'5 84-15681
ISBN 0-470-20096-0

Printed in Singapore by Singapore National Printers (Pte) Ltd.

PREFACE

This manual of geology has been developed over a decade of teaching geology and engineering geology to undergraduate civil engineering students in the United Kingdom and Singapore. During that time, the course has changed dramatically, and I hope has improved so that it now meets the requirements of students aspiring to become professional civil engineers. One major lesson which has been learned over that decade is that students are more interested in subjects which they find relevant to their chosen vocation. It has therefore been essential to couch geology in terms suitable for civil engineers and deal with topics in geology immediately applicable to their needs. Many aspects of geology of a more esoteric nature are indeed fascinating, but within the time constraints of most civil engineering courses would probably be an unaffordable luxury.

As well as attempting to serve the needs of undergraduate civil engineering students, the aim has also been to put sufficient information in the book for it to serve as a useful handbook of geology for qualified civil engineers. To this end, geological methods employed in current civil engineering practice have been covered in some detail. This information should provide a sound base for geology to be used as a kind of materials science for geotechnics. Without some understanding of the geological make-up of soils and rocks both in the mass and in sample form, soil mechanics and rock mechanics can never fully be appreciated.

The geological information required by civil engineers is however not the same all over the world. The raw materials of soils engineering in particular are very different in the northern latitudes from those in the tropics. Relatively little coverage has been given to tropical conditions in the past, a deficiency which this book tries to remedy.

I should like to thank Mr H Dutton and Dr T Cairney of the Department of Building and Civil Engineering, Liverpool Polytechnic, UK, and Professor C N Chen of the School of Civil and Structural Engineering at the Nanyang Technological Institute, Singapore for their encouragement, and for recognizing the importance of geological training for civil engineers.

Thanks are also due to Yeo Chin Soon for preparing the many illustrations in the book and Jamillah Sa'adon for typing the manuscript.

June 1984

John Pitts
Singapore

ACKNOWLEDGEMENTS

The publisher and author would like to thank the following for permission to quote or modify copyright material (figure numbers refer to this publication, full citations can be found by consulting captions and references).

American Association of Geographers for Fig. 4.1; American Geophysical Union for Fig. 4.17; Edward Arnold (Publishers) Ltd. for Figs. 7.6, 7.7, 7.8, 7.9, 7.13, 7.14; Butterworths for Table 5.2; W.H. Freeman and Company for Fig. 4.18; Geological Society of London for Figs. 1.3, 4.11, 4.12, 4.13, 4.14, 4.15, 6.20, 6.21, 6.22, 6.23, 6.24, 6.25, 6.26, 6.27, 7.1, 7.2, 7.3; The Geologists' Association for Fig. 1.1; The Institution of Civil Engineers for Fig. 1.4; The Institution of Mining and Metallurgy for Figs. 2.22, 3.11, 3.12, 3.13, 3.14, 3.15, 3.16, 3.18, 7.10, 7.11, 7.15, 7.16, 7.17, 7.18, 7.19; International Association of Engineering Geology for Fig. 6.19; Longman for Figs. 4.2, 4.4; National Research Council of America for Figs. 4.16(a) to (j) inclusive; Norwegian Geotechnical Institute for Figs. 3.5, 3.17; The Open University for Figs. 6.1, 6.2, 6.3; Singapore Journal of Tropical Geography for Fig. 4.19.

CONTENTS

A MANUAL OF GEOLOGY FOR CIVIL ENGINEERS

CHAPTER 1: BASIC CONCEPTS IN GEOLOGY AND THEIR RELEVANCE IN CIVIL ENGINEERING

1.1 Plate Tectonics

This is probably the most important theory in geology, as most of the processes and products seen at the earth's surface may be related to this. It has been found that the earth's crust is divided up into a series of large slabs called plates and that these are continually moving relative to each other. The movement is essentially determined by the processes occurring at the margins of the plates. At CONSTRUCTIVE margins, new material is reaching the surface of the earth from great depths within the earth's interior. This almost always occurs in mid-ocean and forms large submarine volcanic ridges like the mid-Atlantic Ridge. The rising of the magma (molten rock) to the surface is believed to occur because of the existence of massive convection currents in a semi-molten zone beneath the crust, called the mantle.

As new material is erupted at the surface of the earth, it is added on to the rear edge of the two plates forming the constructive margin (Fig. 1.1).

At the opposite side of each of the plates is a DESTRUCTIVE margin, where crustal rocks are being recycled by "subduction", that is, plunging down into the depths of the earth to be remelted, and perhaps recirculated within the mantle. Subduction mainly occurs of the dark, heavy rocks

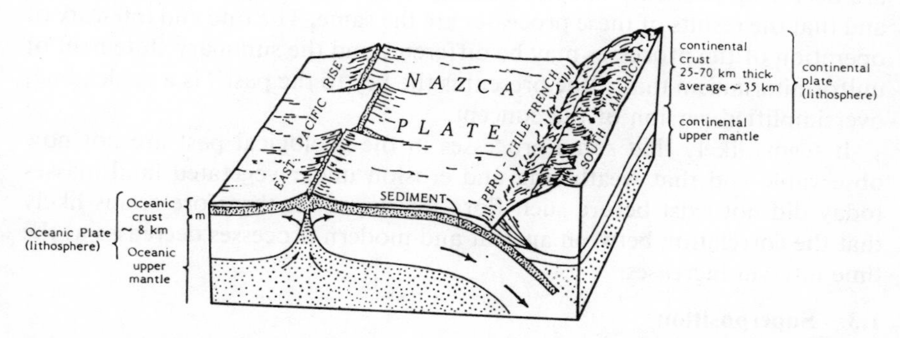

Fig. 1.1 Plate tectonics — constructive and destructive margins. (Oxburgh, R.E., 1974; Copyright © 1974 by *The Geologists' Association*)

which were erupted hundreds of millions of years before at the constructive side of the plate. The reworked silica-rich rocks are less dense and form continents. They "ride" on top of the denser rocks and are crumpled as one dense, subducting plate, often forming the ocean floor runs into a less dense over-riding plate. This results in mountain chains like the Andes, Rockies, and Alps. The mountains are formed from crumpled sediments and remelted rocks often producing volcanic activity. The subduction zones may be adjacent and approximately parallel to the coast of a continent, as with the west coast of South America, or may be some distance offshore producing a line of volcanic islands called an Island Arc, and typified by the Japanese, Indonesian and Philippine Islands.

At destructive margins, the disturbances caused by subduction resulting from the frictional resistance and break-up of the downward plunging plate, leads to the generation of earthquakes.

In some areas, the plates are running alongside each other rather than against each other. This produces large breaks or fractures in the crust called faults such as the San Andreas fault in California. This is known as a CONSERVATIVE margin and during sudden periods of fault movement, earthquakes are generated, e.g. San Francisco in 1906.

So, between constructive, destructive and conservative margins, we are able to account for most of the major geological processes and features of the earth: the distribution of continents and oceans; the latitudinal position of continents and the types of rocks being formed there; major mountain ranges; areas of volcanic activity; earthquake zones; origin of tectonic processes; metamorphic belts; major tear faults; etc. The whole concept unifies a large number of geological observations, and is widely used to explain both recent and more remote events in earth history (Oxburgh, 1974).

1.2 Uniformitarianism

This concept, outlined in the very earliest years of geology by James Hutton is fundamental to the subject. It maintains that the processes which are occurring presently are the same as those which operated in the past, and that the results of these processes are the same. The rate and intensity of operation of the processes may be different, and the summary statement of uniformitarianism that, "the present is the key to the past" is a misleading, oversimplified version of the concept.

It seems likely that some processes in the geological past are not now observable and that weathering and erosion under vegetated land masses today did not exist before such flora developed. It therefore seems likely that the correlation between ancient and modern processes decreases as the time interval increases.

1.3 Superposition

This is perhaps the earliest of all geological concepts and states that in a series of rock strata, the upper members of the series were formed after the lower members. Only in areas of considerable tectonic deformation where beds have been inverted, is this not true, and in such situations, certain features of the rocks may assist in determining their correct "way-up".

1.4 The Geochemical Cycle

The wide ranging series of geological processes and products may conveniently be presented as a cycle of events (Fig. 1.2). The course taken by primary materials, molten rock derived from a position beneath the earth's crust and known as the mantle, can be charted. Its modification on reaching the surface, the variety of processes influencing it, and the new rocks derived from it, can be shown by a series of linked processes.

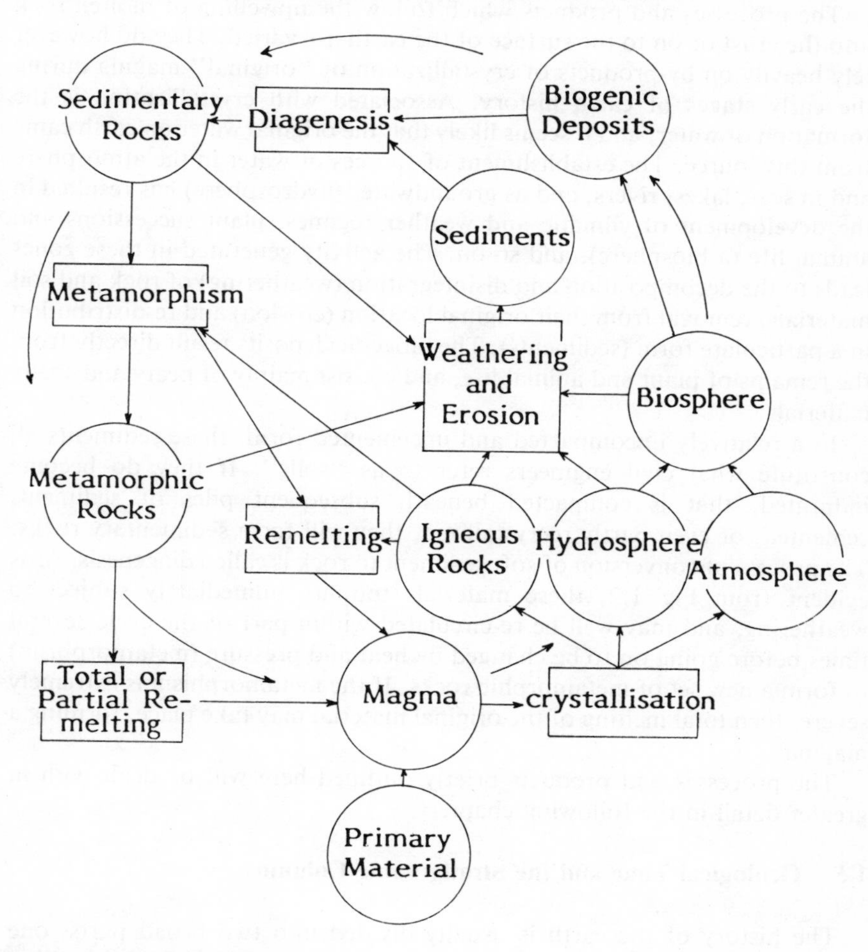

Fig. 1.2 The Geochemical Cycle.

Of course, in the field, we see the products of the partially completed cycle, and during the course of the cycle, some parts may be repeated several times before the cycle is completed. The time taken to complete the cycle is obviously variable, but is likely to be measured at least in hundreds of millions of years.

The course of the geochemical cycle is ultimately related to the cycle of plate tectonic movements previously discussed. Primary material is most obviously represented by the upwelling of submarine lavas at constructive plate margins. These materials then enter the tectonic plate ''conveyor'' system and eventually, having undergone much modification, approach the destructive margin where they are likely to be metamorphosed and re-melted to magma. Some may be retransported to the mantle area to take their place once more as ''primary material''.

The processes and products which follow the upwelling of molten rock into the crust or on to the surface of the earth are varied. They do however rely heavily on by-products of crystallization of ''original'' magma during the early stages of earth history. Associated with crystallization is the formation of water, and it seems likely that the original water on earth came from this source. The establishment of sources of water in the atmosphere and in seas, lakes, rivers, and as groundwater (hydrosphere) has resulted in the development of climatic and weather regimes, plant successions and animal life (a biosphere), and so on. The activity generated in these zones leads to the decomposition and disintegration (weathering) of rock and soil materials, removal from their original location (erosion) and re-distribution in a particulate form (sediments). The biogenic deposits result directly from the remains of plant and animal life, and consist mainly of peaty and shelly materials.

In a relatively uncompacted and uncemented form, these sediments all constitute what civil engineers refer to as ''soils''. If they do become indurated, that is compacted beneath subsequent piles of sediment, cemented, or even partly recrystallized, they will form sedimentary rocks. The process of conversion of soft sediment to rock is called diagenesis. As is evident from Fig. 1.2, these materials too are immediately subject to weathering, and may well be re-circulated within part of the cycle several times before going on to be changed by heat and pressure (metamorphism) to form a new set of metamorphic rocks. If the metamorphism is extremely severe, then total melting of the original material may take place, forming a magma.

The processes and products briefly outlined here will be dealt with in greater detail in the following chapters.

1.5 Geological Time and the Stratigraphic Column

The history of the earth is usually divided into two broad parts, one called the PRE-CAMBRIAN which covers approximately the first 4000 million years, and the PHANEROZOIC which deals with the last 600 million years. The Phanerozoic in turn is divided into three eras, PALAEOZOIC (ancient life), MESOZOIC (intermediate life), and CAINOZOIC (new life). So, most of the Phanerozoic has been divided up on the basis of fossils found in rocks. Although some plants and animals did exist before that time, they were generally very primitive, and did not leave many "hard parts" to become fossilized. Only after the Pre-Cambrian did shelly, marine animals begin to flourish and diversify. Eras consist of several periods which in turn may be divided into Lower, Middle and Upper parts. Conventionally, when writing or drawing a geological succession, the oldest formation goes at the bottom of the page, thus obeying the law of superposition.

Most of the PERIOD names are derived from areas where rocks of that age form a near-complete succession, have been studied in detail, and are "typical". These are called "type areas". Most of them are in Great Britain and Western Europe. Their use, with very few exceptions is worldwide. Only the finest sub-divisions of the stratigraphic column relate to local areas, and the same principle is used.

In the case of igneous rocks, these do not of course contain fossils, and the ages of these rocks are determined by dating of radioactive elements contained within some of the minerals.

1.6 Geomorphic Systems

Geomorphology is the study of the form of the earth's surface and the processes which sculpture it. Systems theory is to do with the study of a series of interrelated features in a landscape and the ways in which they react with each other (Chorley, 1962). So, the study entails the mass and energy of a geomorphic system, how much enters (input), is transferred from point to point within a system (throughput), and leaves a system (output). Any system tries to create a balance between the three components, an example of geological equilibrium. This is often achieved over a short period of time, but the balance rarely survives major disturbances like a severe storm, an earthquake, or the interference of man. The last of these tends to produce permanent changes within a system and any return to equilibrium conditions within a geomorphic system may be along rather different lines.

Drainage basins are regarded as fundamental landscape units in geomorphology and can readily be dealt with as systems. However other features of landscape frequently encountered in civil engineering activity may also be dealt with as systems. Slopes are a notable example (Brunsden, 1973), and this concept was applied to a part of Western England where designs were required for coast protection works (Pitts, 1983a). The various facets of the system are shown in Fig. 1.3. The slopes were in balance for most years, and the material slipping down the face to the beach was removed by wave action without major erosion of the in situ material at the toe of the slope.

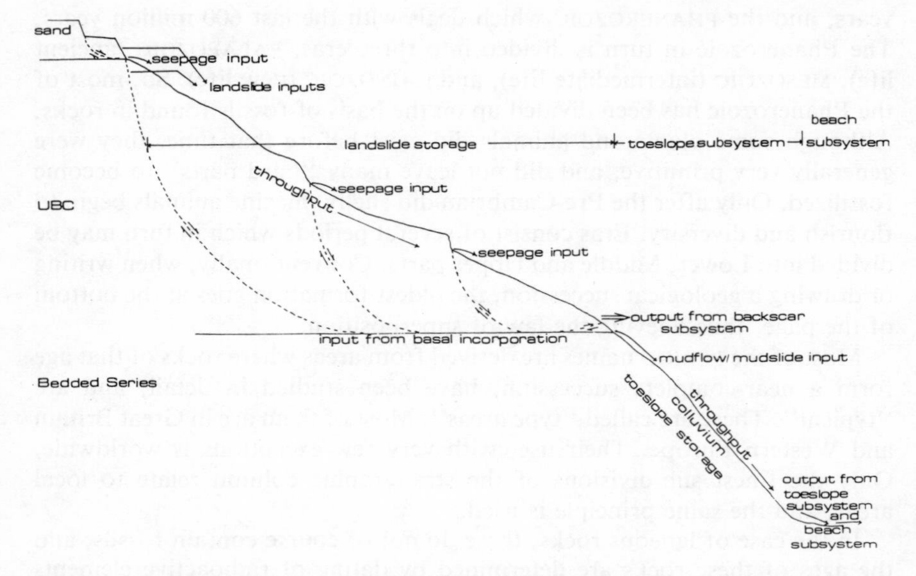

Fig. 1.3 Systems theory — the elements of a landslide system on a sea cliff. (Pitts, J., 1983a; Copyright © 1983 by *Geological Society of London*)

The nature of the slope instability could also be determined from close scrutiny of the interrelationships between the form of the slope, the geological succession at the site, and seepage. This enabled a slope budget to be established, also shown in Fig. 1.3.

1.7 Thresholds and Uniformity of Natural Systems

Natural physical systems operate under natural physical laws, although the time element involved is complex, varying between uniformity (more or less, slow steady rates), and catastrophe. The dividing line between these is called a THRESHOLD for a system, and may be defined as a condition at which a process or system changes, sometimes abruptly, from a relatively simple to a relatively complex thing; from a relatively predictable to a relatively non-predictable system; from a measurable to a non-measurable system or one measured only with difficulty; from a reversible and controllable to an irreversible and non-controllable state — i.e. from a uniform to a non-uniform state.

Example 1 — shearing strength of materials

Physically, beyond the shearing strength threshold, rocks lose cohesion, and potential energy (stored elastic energy) is converted to kinetic (particle motion) energy.

In the case of large forces and/or large masses, e.g. earthquakes and landslides, the threshold represents a point beyond which strain becomes rapid and relatively unpredictable, irreversible, and certainly uncontrollable. In the case of landslides, additional support given to the slope will in effect increase the threshold of that slope. Although not impossible, this is more difficult to undertake for earthquakes.

Example 2 — stream flow

At a velocity threshold in the channel, flow changes from laminar to turbulent. At a discharge threshold, overbank flow takes over from channel or in-banks flow, i.e. becomes an irreversible and unmanageable system. Flood prevention is mainly aimed at keeping a stream within its banks, e.g. by dykes or levees, and this way of extending the threshold is generally possible in a way not so for earthquakes.

So, dynamic geological processes operate within thresholds. Beyond these they leave their steady-state condition. Control is by keeping a process below its threshold, extending the threshold, or providing an adequate monitoring and warning system. The key lies in being able to define or estimate thresholds for given geological processes in particular environments as they are likely to affect civil engineering activities. This is why both laboratory and field tests on representative samples are so important in geotechnical engineering. Shear strength, modulus of elasticity, pore-water pressures, etc. are all thresholds for which values must be known if designs are to be realistic and safe.

1.8 Magnitude and Frequency of Forces in Geomorphic Processes

This concept (Wolman and Miller, 1960) is based on the facts that landscape sculpture depends mainly on:

(a) Variable atmospheric processes
(b) Gravity

To be effective, a "threshold" for any process must be exceeded, and beyond this there occurs a wide range of *magnitudes* of forces. The problem is to assess the relative effectiveness of, on one hand catastrophic events, and on the other, more ordinary events on landscape evolution.

The relative importance can be measured in terms of the relative amount of work done on the landscape.

When annual floods are known for a series of years, *frequency-analyses* may be carried out. Such analyses are concerned with *magnitude* (amount of discharge) and with *frequency*. The *recurrence interval* (or *return period*) of a flood of given magnitude is expressed in years.

For example, the flood which at a particular station has a recurrence-interval of 25 years is called the 25-year flood. This flood can be expected, as an annual peak, once during a 25-year period. Likewise, the 100-year flood can be expected once in a century.

This does not mean that the 25-year flood can be expected at intervals of 25 years. It is possible for a station to record the 25-year flood for the period 1941–1965 in 1965, and the 25-year flood for 1966-1990 in 1966. Although it would be unusual, it is not impossible for two 100-year floods to occur in successive years; also, damage by two successive 100-year floods could be followed, in a third year, by even more severe damage from a 1000-year flood. Very wet years often occur in groups; this results in bunching.

In the study of rivers, the largest proportion of material is carried by flows occurring on average between one and several times in 1–2 years. However, as the variability of flow increases (deserts), and the size of the drainage basin decreases, a larger percentage of the load is carried by less frequent events.

In many basins, 90% of the material transported from the basin as a whole, is removed by storm discharges which occur at least once every 5 years.

The amount of work done by an event is not necessarily the same as its relative importance in landscape sculpture. It is measurable both by magnitude and frequency. Most rivers in flood carry a vast amount of sediment. The relative importance is evaluated by comparing the quantities carried in a rare vast flood to that in a more frequently occurring flood of smaller magnitude. The largest amount of material is eroded from drainage basins by small or moderate floods. Vast floods carry vast quantities of sediment, but occur so rarely, that from the standpoint of transportation, their effectiveness is less than for smaller, more frequent floods. 150 river basins in the USA were studied, and events occurring at least once in 1–2 years, and usually occurring several times, did most transporting of sediment.

Analysis of return periods for major floods is of great significance in dam and spillway design.

Beaches develop towards an equilibrium profile which may be considered the average state, but one which undergoes frequent fluctuations. Storm waves may periodically destroy the equilibrium profile, but over a number of years, an average profile will develop by which a beach may be characterized, i.e. the effective stress to which the equilibrium profile of a beach is related is one produced by moderate storm waves, rather than by waves which accompany severe storms.

Catastrophic events produce results which are, in some respects, unique. Landslides commonly occur with extreme downpours, as was the case in western Singapore following high rainfall during November 1982 (Pitts, 1983b). New gullies also form during extreme precipitation and once initiated, the gully will further grow under moderate rainfall events. In Andean South America, avalanches generated by thaw or by earthquakes involve vast volumes of rock moving very fast and scouring everything off the landscape.

Severe erosion followed an earthquake of magnitude 7 in Papua New Guinea. The earthquake induced slope failures over an area of 240 km^2. Approximately 60 km^2 of steep slopes were stripped of soil and vegetation by debris avalanches, transferring in a few seconds, this material to the river channels. A total of 27.6×10^6 m^3 of material was removed as a result of the earthquake and of this, half entered the sea within the first 6 months.

In this area of the Adelbert Range, the rate at which the landscape is worn down is estimated as averaging 80–100 cm/1000 years. Of this, 60–70% can be attributed to earthquakes. This is typical for similar areas within the humid tropics, but much higher than average rates for areas outside the humid tropics.

Changes in the dimensions and positions of stream channels frequently occur in large floods. In the winter of 1969–70, Tunisia experienced its first significant rainfall for 9 years, resulting in widespread sheet floods over a normally fairly stable landscape, largely devoid of vegetation. Little percolation occurred because of the relatively large amount of clay in the soil. In tropical forest areas, large magnitude erosive events may be generated by man's influence. These areas normally have a deep weathering profile, but very low suspended and dissolved sediment loads in comparison. Once vegetation is cleared though, the suspended matter increases greatly. In many tropical catchment areas, it can be shown that in headwater regions, where man's influence is minimal, erosion is far less than downstream, where the effects of urbanization, cultivation, deforestation, construction and mining, lead to large changes in hydrological, sediment and morphological characteristics. The effects of urbanization are summarized in Table 1.1.

Extreme floods accomplish transport of material that is impossible under normal flow, and in general, as far as sediment size transported is concerned, the effects of floods appear to be directly proportional to their magnitude.

In coastal areas, where marine cliffs are fronted by broad beaches, erosion may only occur on these cliffs during periods of extreme wave action. In situ soils and rocks at the toe of a cliff may periodically be eroded during severe storms by waves of a much greater magnitude than normal. This tends to lead to an acceleration of slope processes behind, generally disturbing the established balanced state where outputs by erosion become greater than inputs. However, once the storm is passed, processes will again tend to be of a moderate magnitude and frequency and will work towards regaining a balanced state between input, throughput and output on the slope. The increased erosion at the toe will however have resulted in a large increase in input at the top of the slope and redistribution of this large new supply will take some considerable time, thus delaying the re-establishment of the "normal" balance. Barrier islands which are stable

Table 1.1 Changes in the nature and behaviour of channels as a result of urbanization in their drainage basins.

A. Hydrological changes

(a) Modifications in the flow/duration curve

(i) lower base flows
(ii) larger and quicker floods

(b) Changes in flood frequency

(i) increases in the frequency of floods of a given size
(ii) increases in the size of a given flood for a particular return period, e.g. mean annual flood

(c) Modification of runoff from individual storms

(i) decrease in lag time between rainfall and stream flow
(ii) increase in size of flood peaks
(iii) increase in peak velocity

(d) Changes in water quality

B. Changes in sediment characteristics

(i) deposition of relatively coarse sediment in channels as a result of construction in the early stages of urbanization

(ii) reduction in delivery of sediment to channels at later stages as solid urban surfaces are established

(iii) new set of depositional features in channels

C. Morphological changes

(i) changes in channel size, width and depth
(ii) changes in depositional features in channels

(Modified from Rahman, A and Gupta, A (1980))

under ordinary conditions may be eroded below mean tide level during extreme storms. Similarly, along shores of low coastal plains, beach ridges may be built several feet above mean low water during infrequent events of great magnitude.

Evaluation and relative effectiveness of various geomorphic processes in a given region, as well as relative effectiveness of different frequencies will require more detailed observation on both processes and forms. Only then will predictions for long-term planning and design become more reliable.

1.9 Equilibrium of Geological Processes and their Disturbance by Engineering Activity

Balance in geological and geomorphic systems is a delicate condition which takes long periods of time to establish. The balance is easily disturbed, by among others, civil engineering activity. It is of great importance to try to understand how geological systems and processes influencing those systems are likely to react to disturbance. Several examples are briefly covered to demonstrate some examples of such disturbance.

Example 1

Levels of in situ stress in geological materials result from a more or less complex array of geological processes (see Chapter 3). These range from overburden to episodes of tectonic deformation. The stresses may be disturbed in a number of ways by construction activity and processes permitting the redistribution of stresses often lead to situations which require careful control and monitoring.

(a) *Settlements under increased load*. When structures are placed on the ground, settlement of some magnitude is likely to occur as a result of the additional vertical stresses which result from the weight of the structure. The most obvious example is that of the consolidation of a clay layer beneath a foundation. The additional applied stress leads to a slow squeezing out of water from pores in the clay, a closing of the pores, and a decrease in the thickness of the clay layer. This results in a settling down of the structure.

(b) *Generation of critical pore pressures under additional loads*. This situation is closely related to the consolidation of a clay layer. The additional stresses which result from construction are partly taken up by the pore water pressures in the soil. These have the effect of pushing apart the grains. This generally decreases the strength of the soil by reducing the effective normal stress due to the overburden. As long as the pore pressures are not so high that they become critical for the strength of the soil, they will gradually dissipate under the new load conditions at a rate in keeping with the permeability of the soil. It is the low permeability of clays which makes them unable to adjust the pore pressures quickly to new load conditions. When an embankment is built, the rate of loading either has to be strictly controlled to allow dissipation of pore pressures from the first stage of construction, or, more likely, additional drainage is provided for a clay layer to accelerate the pore pressure dissipation and the gain in strength.

(c) *Vertical relief of in situ stresses*. In some areas of strong tectonic deformation, high horizontal stresses result in significant vertical stresses which may be relieved in the floors of excavations. During the construction of the Hoover Dam in the USA, good rock conditions could not be obtained in the floor of the main dam excavation because of stress relief and break-up of the rock. As this disintegrated rock was removed, the material beneath became less confined and in turn started to break up. The conclusion was that the rock mass required treatment before construction could commence.

(d) *Lateral relief of in situ stresses*. This tends to occur on slopes and in tunnels. One particular example of stress relief in slopes demonstrates that problems of stress relief exist in soil as well as rock masses. In overconsolidated fissured clays, excavating cuttings leads to stress relief and an opening of small planes of weakness (fissures) which cut through the clay. This starts a long term weakening of the soil which may result in failure of the cutting several decades later (Skempton, 1964, 1970). In tunnels, explosive stress relief may result. In deep tunnels beneath mountains where in situ stresses are very high, large slabs of rock may fly off into the tunnel workings after excavation has provided space into which the pent-up stresses may be relieved.

(e) *Changes in stress by the introduction of water.* Groundwater generates pressures which increase the shearing stresses acting upon soils and rocks. Increasing hydrostatic pressures during engineering activity may have widespread effects. Impounding of a reservoir results in a raising of groundwater pressures to new equilibrium conditions. This may increase shear stresses in rocks and soils sufficiently to cause landslides (e.g. Vaiont in northern Italy in 1963), or along major breaks in the ground (faults) sufficient to generate earthquakes. Seismic events have occasionally been noted as resulting from impounding reservoirs in certain areas where earthquakes were previously unknown.

Example 2

Groundwater regimes exist in equilibrium states, although that does not necessarily mean at uniform levels. Seasonal variations and tidal effects are two causes of relatively short term variations which may be experienced in groundwater levels. These may have observable results, for example in Scandinavia where the groundwater levels rise during the rainiest month and in the month of thawing of snow and ice (Terzaghi, 1962). These two months correspond with the periods of greatest landslide activity (Fig. 1.4).

(a) The flow of water within the ground is part of the regime. It can be diverted quite easily by tunnelling activity. The tunnel is a low pressure area within the ground, and since groundwater flows from areas of high to areas of low hydrostatic pressure, then the tunnel acts as a large drain. In addition, the effects of stress relief causing an opening of the cracks in the rock or even the propagation of new ones, will in turn increase significantly the water-holding capacity of the ground. Hence, groundwater flow into tunnel workings is a common problem and compressed air or slurries are used to combat the effects of this.

(b) When a reservoir is impounded, a great deal of leakage takes place into the soil and rocks around the reservoir. This should be a short-term effect at least on a large scale, but is essential in raising regional groundwater tables to levels in equilibrium with the top water level of the reservoir itself. If this does not occur, then serious losses from the reservoir may make the project uneconomic. Directions of ground-

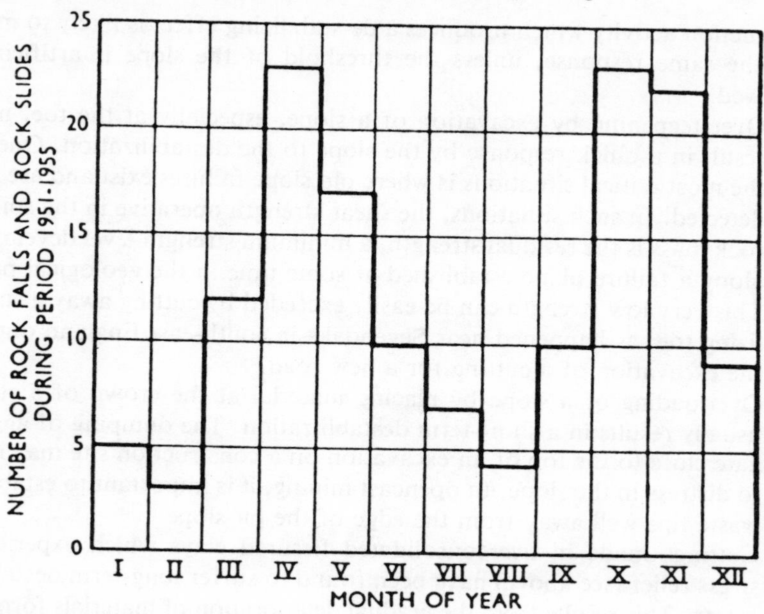

Fig. 1.4 Distribution of rockfalls and rockslides in Norway with time. (Terzaghi, K., 1962; Copyright © 1962 by *Institution of Civil Engineers*)

water movement may also change as a result of these adjustments, and long-term leakage is likely to result in a series of new seepage points or spring lines downstream of the reservoir or even in adjacent valleys. This is often accompanied by an increase in landslide activity.

(c) Established groundwater levels may need to be lowered for excavation to take place safely during civil engineering works. By doing this, soil pores previously filled with water become filled with air, and hence compressible. The result may be settlement of existing structures close to the excavations.

Example 3

Natural slopes and cut slopes have produced severe civil engineering problems in the past. Natural erosional slopes try to attain profiles which are in equilibrium with prevailing conditions. That may result in: (i) vertical cliffs where severe erosion is maintained at the toe; (ii) slopes of medium inclination (the gradient largely depending on the materials forming the slopes) where a balance between erosion on the slope and removal of material from the toe is maintained; (iii) slopes of low gradient, where no erosion is occurring at the toe, and where shallow movements in the surface layers are taking the slope towards its angle of ultimate stability. Unstable slopes adjust by failing, which has the result of lengthening the slope, decreasing its gradient, and redistributing stresses acting on the soil or rock mass. Any construction activity which produces a de-stabilizing effect is likely to meet with the same response, unless the threshold of the slope is artificially improved.

(a) Oversteepening by excavation of a slope, especially at the toe, may result in a quick response by the slope to the destabilization. One of the most critical situations is where old slope failures exist and are not detected. In such situations, the shear strength operative in the soil or rock mass is the residual strength, a minimum strength level developed along a failure plane established at some time in the geological past. This very low strength can be easily exceeded by cutting away a stabilizing toe, as happened near Sevenoaks in south-east England during the excavation of a cutting for a new road.

(b) Overloading of a slope by placing material at the crown of a slope usually results in a short-term destabilization. The dumping of aggregate close to the top of an excavation on a construction site may lead to distress in the slope. In opencast mining, it is important to establish waste tips well away from the edge of the pit slope.

(c) Cuttings made in overconsolidated fissured clays which experience stress relief (see above) have been found to suffer long-term destabilization. This results from the gradual deterioration of materials forming the slope by softening during water ingress, by increases in pore-pressures as equilibrium moisture contents rise and by shear creep along the fissure surfaces which act as local concentrators of stress.

Example 4

Rivers tend to produce carefully balanced systems, producing ultimately, what is referred to as a graded profile. Each part of the river tries to attain balance with all other parts, and a disturbance at any point on the river may cause serious reactions elsewhere.

(a) Rivers grade to a base level, usually the sea, or even a lake. When a lake level is raised to form a reservoir or a new reservoir is developed, the profile of the stream must adjust so that it flows into the lake at the "correct" level and so that the profile behind is at a suitable series of gradients. This is likely to entail deposition of large amounts of materials in the channel which may have an influence on channel stability, propensity to flood, etc.

(b) The energy in rivers is normally expended on the transport of water and sediment. Downstream of a dam, mostly clear water will be in the channel, 95–99% of the normal sediment of the river being retained in the reservoir. This results in accelerated erosion of the bed and banks downstream of the dam in order to obtain a new sediment supply. The increase in erosion may have implications for structures built-in or adjacent to the river. Significant proportions of sediment can only pass the dam with the installation of costly devices.

(c) Patterns of erosion may be changed in a number of ways by introducing defense and training schemes, bridge piers, temporary works, etc. In general, river taining results in straightening the course of a river and decreasing the roughness of its wetted perimeter. This is likely to allow greater erosion or transportation to occur since less energy losses would accrue overcoming friction in the channel. Bridge piers deflect and channel water and frequently cause scour upstream of the piers where the water builds up behind the constricted part of the channel. This can then erode the banks around bridge abutments. Many effects may result from this and model studies are often undertaken to investigate changes in flow patterns which result from placing structures in a channel.

Temporary works may also cause problems of accelerated erosion. During the laying of new sewer pipes for Merthyr Tydfyl in South Wales, a coffer dam was constructed in the middle of the River Taff. This effectively concentrated the flow of the river into half of its normal channel width, and increased significantly its erosive potential. The undercutting of the steep banks of the river at the foot of a 30–metre slope caused a series of slope failures on the slope. These resulted in the loss of part of the line of a proposed new road, which subsequently had to be re-routed.

Example 5

Beaches possess equilibrium profiles, although these may be seasonal. On the east coast of England at Scarborough, the prevailing winds during summer are south-westerly and result in light seas and deposition on the beach. During winter, the winds shift to a north-easterly direction, drive larger eroding waves onshore, and result in a scoured, lower equilibrium profile.

The sediment supply along a coast is also delicately balanced. Any interference with this may have profound effects at distant localities which rely on the along-shore transport of material by waves and currents.

(a) Shore defense works are constructed to solve local problems. Coastal systems however are often extensive, and the results of interference further along the coast may be considerable. Natural beaches form excellent defenses, but the supply of sediment to a beach must be maintained. If at one locality groynes or similar beach retaining structures are built, they will intercept the sand moving along the coast upon which the stability of beaches elsewhere rely. Once starved of their supply, the beach profile will become lower and increased erosion of the backing coastline may result.

At Caldy on the Dee Estuary in western England, large boulders were placed at the foot of a cliff which was suffering considerable erosion. The effect of this protection was to cause scour of the beach at an adjacent section, and thereby to increase the erosive power of the waves. Severe undercutting and instability of the cliffs resulted over a distance of 120 metres (Pitts, 1983a).

Scour of the foreshore has also occurred in front of sea walls. A variety of wall profiles has been developed, but the height of the water after impact of a wave against the wall concentrates considerable erosive power in front of the wall. Beach lowering of 1–2 metres has occurred at some points on the north foreshore at Scarborough in eastern England, resulting in exposure of the foundations of the sea wall.

This wide variety of examples is meant to show the relevance of geological and geomorphological concepts to civil engineering works. Equilibrium conditions in geology and their vulnerability are important considerations, and much time, trouble and money can be saved if an appreciation of them is gained before rather than after construction.

CHAPTER 2: ROCKS, THEIR COMPOSITION, IDENTIFICATION AND PROPERTIES

Rocks of all kinds are made up of an assemblage of minerals, and most of the minerals, of which there are a vast number, have strange names. It is sufficient that only a small number of the commonest rock-forming minerals will be introduced to aid with rock identification. The engineering properties of the rocks often relate to the minerals which make up the rock, and their arrangement. The minerals may not be the original ones which formed the rock, and some of these secondary minerals are also introduced.

Most of the mineral names will be unfamiliar and therefore to keep new names to a minimum, mineral groups are mainly presented with only the commonest of their members individually dealt with. Tables 2.1, 2.2 and 2.3 are summaries of the characteristics of the most common minerals. On some occasions, there may be a need to know others, so these tables are not meant to be exhaustive.

The most important of the physical and chemical properties of these minerals are to be used to aid with their identification.

For civil engineering purposes, there are several other features of rocks in addition to mineral content which will help with identification. It is really only in igneous rocks that details of the mineralogy are required to accurately identify the rock. For engineering purposes, such accurate identification is not really required. Simplified criteria such as the relative proportions of dark and light minerals and presence or absence of quartz, which is easily recognized, are sufficient. That information along with details of texture, and crystal size will then enable a satisfactory identification to be made.

2.1 Igneous Rocks

Igneous rocks are normally crystalline in nature, having solidified from a silicate melt (magma). They may be extrusive, i.e. crystallized at the earth's surface (volcanic), or intrusive, where they originate at depth and cool more slowly (plutonic). Extrusive rocks are fine-grained to glassy, whereas intrusive rocks are medium to coarse-grained, largely depending on the size of the intrusion.

Table 2.1 Main minerals in igneous rocks — physical characteristics.

	MINERALS AND COMPOSITION	lustre	hardness	cleavage	twinning	colour	streak	fracture	specific gravity
	QUARTZ SiO_2	vitreous	7	none		colourless — white (but for impurities)	white	conchoidal	2.65
FELDSPARS	ORTHOCLASE $KAlSi_3O_8$	vitreous/ pearly	6	two, good approx. 90°	simple	white, greenish, pink, grey	white	conchoidal, uneven	2.56
FELDSPARS	PLAGIOCLASE $(Na,Ca)(Al,Si)_4O_8$	vitreous	6	two, good approx. 90°	compound	white, grey	white	uneven	2.7 approx.
MICAS	MUSCOVITE $KAl_2(AlSi_3O_{10})(OH, F)_2$	vitreous/ pearly	2½	one perfect		colourless pale grey, white	white	—	2.85 approx.
MICAS	BIOTITE $K(Mg,Fe)_3(AlSi_3O_{10})(OH,F)_2$	vitreous	2½	one perfect		dark brown black	white	—	3.0
	HORNBLENDE $NaCa_2(Mg,Fe)_4(Al,Fe)(Si,Al)_8O_{22}(OH,F)_2$	vitreous	5½	two, good approx. 60°		dark green black	white, grey	uneven	3.2 approx.
	AUGITE $(Ca,Mg,Fe,Al)_2(Al,Si)_2O_6$	vitreous	6	two, poor approx. 90°		black	white		3.3
	OLIVINE $(Mg,Fe)_2SiO_4$	vitreous	6½	very poor		olive green weathers to brown	colourless	conchoidal	3.4

Table 2.2 Main minerals in igneous rocks — optical characteristics.

	MINERAL	refractive index	colour	relief	idealized shape	cleavage	twin	interference colours	extinction
	QUARTZ	1.55	clear colourless	low	often irregular	none		grey	
FELDSPARS	ORTHOCLASE	1.52–1.54	colourless, often turbid	low		sometimes traces of one or two	none/ simple	grey	generally oblique
FELDSPARS	PLAGIOCLASE	1.52–1.58	colourless often turbid	low		sometimes traces of one or two	compound	grey	generally oblique
MICAS	MUSCOVITE	1.59–1.62	colourless	moderate		one v.good parallel long direction		bright	straight
MICAS	BIOTITE	1.61–1.65	brown-yellow: pleochroism	moderate		one v.good parallel long direction		bright	straight
	HORNBLENDE (Amphibole)	1.65–1.69	green or brown shades: pleochroism	moderate to high		two possible approx. 60°		bright	oblique
	AUGITE (Pyroxene)	1.68–1.72	colourless or v. pale pink, green or brown	high		two possible approx. 90°		bright	oblique
	OLIVINE	1.69–1.73	possibly green/ brown lines from alteration	high	irregular	very rare		bright	

Table 2.3 Some important minerals in sedimentary rocks.

MINERAL AND COMPOSITION	lustre	hardness	cleavage	idealized shape	colour	streak	fracture	specific gravity
CALCITE $CaCO_3$	Vitreous	3	one, perfect		colourless or white; some impurities	white	conchoidal	2.71
DOLOMITE $CaMg(CO_3)_2$	Vitreous dull	3½–4	one, perfect	usually irregular	white, often tinged with yellow, red brown	white	conchoidal, brittle	2.85
GYPSUM $CaSO_4.2H_2O$	Vitreous/ pearly	1½–2	one, perfect		colourless-white; tinted pink, yellow and grey	white		2.3
KAOLINITE $Al_4Si_4O_{10}(OH)_8$	Dull/ earthy	2	one, perfect when detectable		white-greyish	white		2.6
ILLITE $KAl_4(Si_7AlO_{20})(OH)_4$	Dull/ earthy	1–2	One, perfect	as kaolinite	white to pale	white		2.6–2.9
MONTMORILLONITE $Al_4Si_8O_{20}(OH)_4.nH_2O$	Dull/ earthy	1–2	one, perfect		white, yellow or green	white		2–3
PYRITE FeS_2	Metallic	6–6½	Poor		Brass-yellow	Brown black	conchoidal, brittle	5

1 Classification of igneous rocks

For engineering purposes, a simplified classification of igneous rocks is quite adequate, based largely on mineral content and grain size/textural considerations (Fig. 2.1). It is sufficient to be able to place a particular rock type in the correct family of rocks, and this minimizes the need for familiarity with the complex vocabulary of petrology, (the study of rocks).

There are three categories of grain size; fine, with individual components indistinguishable with the naked eye; medium, with visible individual components, but where they are generally too small to identify with the unaided eye; and coarse, where the individual crystals can be seen and identified. These groups would be generally associated with lava flows, dykes and sills, and larger plutonic bodies, respectively. It is the length of cooling time which determines the grain size.

Virtually all the igneous rocks are composed largely of silicate minerals (Fig. 2.1), and any rock type consists of a definite assemblage of minerals, mainly determined by temperature, which can be used to classify the rock. Although through, for example, re-melting, magmas of quite different

rocks became darker and denser →

Grain size	Acid	Intermediate	Basic	Ultrabasic
Coarse (plutonic)	Granite	Diorite	Gabbro	Picrite Peridotite
Medium (minor intrusive)	Quartz-Porphyry, Felsite	Micro-Diorite, Porphyrite	Dolerite	
Fine (volcanic)	Rhyolite	Andesite	Basalt	
Specific Gravity	2.6–2.8	2.8–2.9	2.9–3.1	3.1–3.3

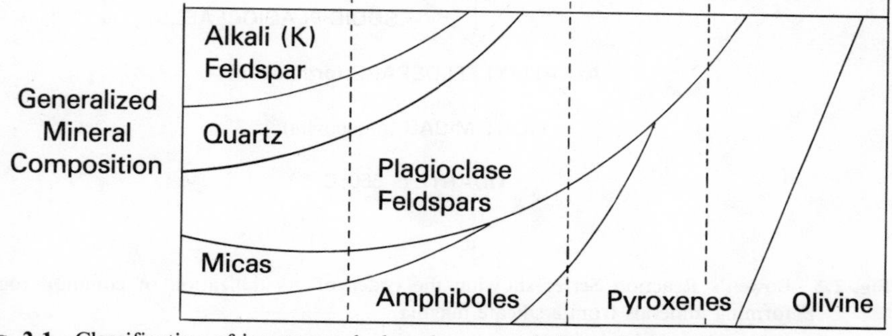

Generalized Mineral Composition

Alkali (K) Feldspar

Quartz

Plagioclase Feldspars

Micas

Amphiboles

Pyroxenes

Olivine

Fig. 2.1 Classification of igneous rocks based on crystal size and mineralogical composition.

compositions exist, it is possible to obtain virtually a full range of rock types from a single magma derived from the mantle. Although six types of igneous rock (basalt, gabbro, andesite, diorite, rhyolite, granite) account for about 90% of all igneous rocks, there is a continuous gradation in the chemical composition of igneous rocks.

By using Bowen's Reaction Series (Fig. 2.2) illustrating the order of crystallization of minerals from a magma, depending on temperature, it becomes apparent how such a varied collection of rocks may be formed from a single source. At the highest temperatures, the rocks formed tend to be free or virtually free of feldspars, containing olivine and pyroxene, and are called ultrabasic. Basic rocks contain a calcium rich plagioclase feldspar, and a pyroxene; intermediate rocks, a plagioclase in between calcium and sodium rich and an amphibole, with less than 10% quartz; acid rocks, a sodic plagioclase and/or an alkali feldspar, and more than 10% quartz. So, the minerals which crystallize out at any particular temperature combine to form a representative rock type, leaving the chemical composition of the still molten portion of the magma changed. Hence, as the high temperature minerals become crystallized out, elements such as Mg, Fe and Ca, are taken from the melt, along with relatively small amounts of silica combined with these as a silicate radical. Thus, the melt is enriched in Al, Na and silica, and the lower temperature minerals and hence, rocks are rich in these.

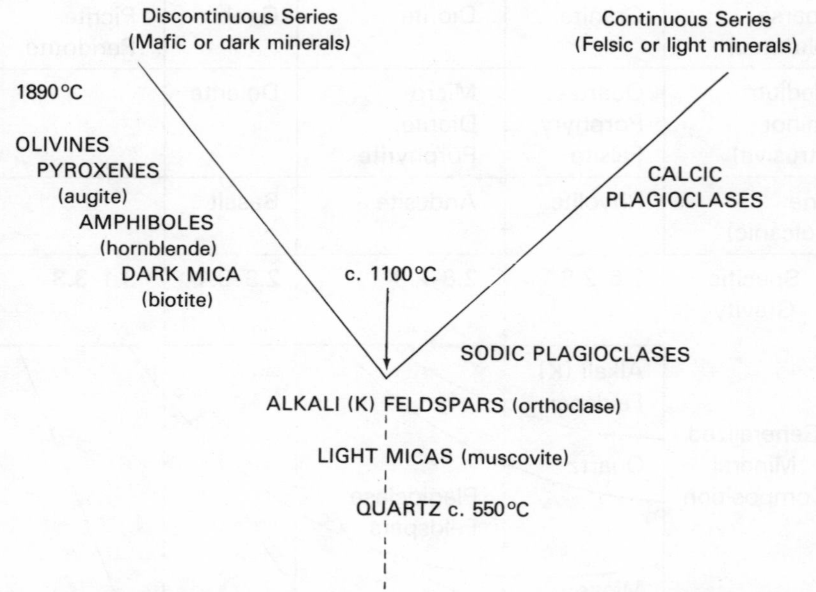

Fig. 2.2 Bowen's Reaction Series showing the order of crystallization of common rock forming minerals from a silicate magma.

2 Control over the form of igneous rocks

Volcanics

(a) Lava Flows

These consist of mobile rock material reaching the surface through vents and fissures and pouring sub-aerially or submarine. The flows to the sub-aerial surface depend for their extent on the fluidity of the magma e.g. Icelandic basalt flows 40 miles long are known.

Central eruptions form volcanoes which also depend for their size and shape on the viscosity of the magma. Basic magmas form huge shield volcanoes, e.g. Moana Loa on Hawaii, which is 100 miles wide at the base and 30,000 feet high from sea floor to summit. The flow rate, which may be 30 m.p.h. leads to widespread sheets. Gas escape is very easy, and explosive eruption is subordinate.

Contrasting with this is eruption of viscous magmas which form smaller steeper sided cones. These lavas tend to have higher silica contents. Rhyolite is the most viscous lava, often forming spines, since the lava can't spread naturally e.g. the spine at Mont Pelee in Martinique which resulted after the eruption of 1902.

There are two general types of lava:

1. Sub-aerial

These lose volatiles quickly due to the rapid decrease in pressure on their arrival at the surface. A crust may form on the top and at the base through chilling, perhaps while the lava is still moving, and results in a tumbled pile of rocks and clinker called blocky lava. Lava retaining most of its volatiles moves more smoothly and without fracture, producing rolls of lava called ropey lava.

2. Submarine lavas

These are characterized by pillow structure, often having glassy margins due to very rapid cooling.

Other structures of lava flows include:

(i) Columnar structure

This occurs in slow cooling heterogeneous rocks and is characteristic of basalt rather than other rocks.

(ii) Amygdales

Basaltic lavas usually lose volatiles to the air and lose fluidity. The trapped bubbles form vesicles, which may be spherical or streaked out. They are often infilled with secondary minerals to form amygdales.

(b) Pyroclastics

Mechanisms forming pyroclastics are probably related to "temperature blocking" in the vent due to cooling and consolidation of a skin of magma during quiet periods. Pressure builds up behind the skin, which may burst leading to fissuring and fragmentation of the rim of the vent. The sudden release of pressure in the magma causes boiling; much gas is released, the magma frothing over and flowing as a fast moving cloud of gas and magma fragments. It settles and is bonded by its own heat.

Intrusives

(a) Sheet forms

Dykes are wall-like bodies, nearly vertical at the time of intrusion, and outcropping as a straight line on a map (Fig. 2.3). They are discordant with the existing structure.

Sills are also tabular, (Fig. 2.3) but horizontal at the time of intrusion. The outcrop follows non-igneous strata, and is frequently concordant with the strata.

(b) Lensoid forms

The relationship depends on the position of the intrusion and the degree of convexity or concavity e.g. Laccoliths (Fig. 2.3) in which overlying strata is updomed the lower strata being near horizontal, e.g. Henry mountains, Utah.

(c) Large intrusions

These are generally acid and are usually referred to as batholiths. The margins of these are usually sharp and steep, but not bottomless. A small batholith is a stock, and a boss is a circular stock in plan.

Batholiths are usually intruded at depth and their exposure normally relies on the erosion of overlying rocks. The rocks into which the hot magma was intruded are usually baked and become contact (in contact with the intrusion) metamorphic rocks. This zone of altered rocks is called an aureole.

2.2 Sedimentary Rocks

Sedimentary rocks are formed at the earth's surface, and therefore form

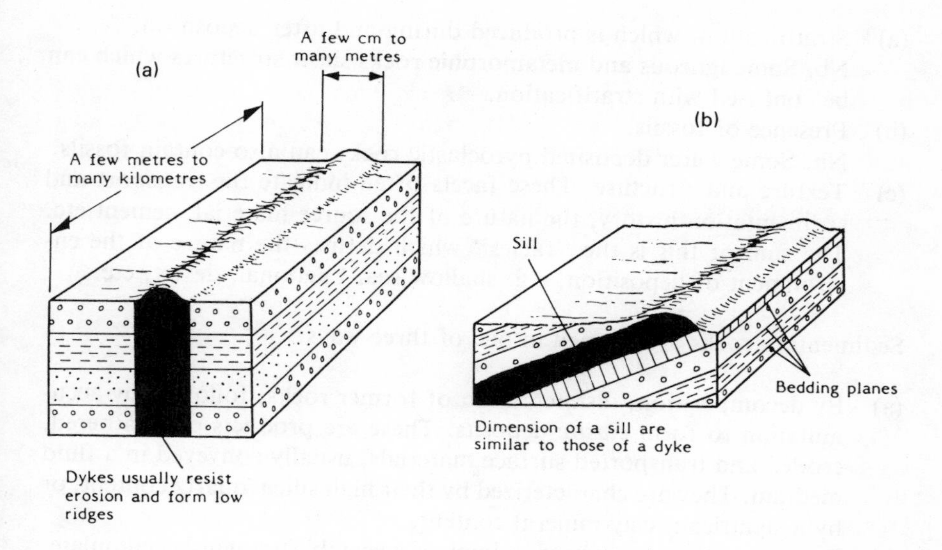

Fig. 2.3 Forms taken by igneous rocks.

under relatively low temperatures and pressures.

Sedimentary rocks are distinguished by:

(a) Stratification, which is produced during and after deposition.
Nb. Some igneous and metamorphic rocks show structures which can be confused with stratification.
(b) Presence of fossils.
Nb. Some water deposited pyroclastic rocks can also contain fossils.
(c) Texture and structure. These facets often indicate the transport and sedimentation history, the nature of the source material, cement etc. The sum of this is the "facies" which defines the nature of the environment of deposition, e.g. shallow sea, Lagoonal, desert, etc.

Sedimentary rocks are formed in one of three ways:

(a) By decomposition, disintegration of former rocks, followed by accumulation to form *clastic* deposits. These are products of weathered, eroded and transported surface materials, usually conveyed in a fluid medium. They are characterized by their high silica/quartz content, or by a significant clay mineral content.
(b) By *organic* means, where animal or vegetable remains accumulate, whether as shell beds, reefs, oozes, etc. They produce carbonate rocks if shelly materials accumulate, or pure siliceous deposits with diatoms (plants).
(c) By *chemical* means. These mainly include evaporites, which are sulphates, silicates, phosphates, chlorates, chlorides, etc. These form mainly by evaporation or precipitation from bodies of surface water.

1 Sedimentation

When materials are deposited in the sea, the heavy particles remain near the shore; the mud is carried well out to sea. The material builds up in layers until the completion of sedimentation, e.g. caused by filling of the basin of deposition.

Sedimentation may also take place in air (desert dunes), or from ice. Deposition largely depends on gravity, and produces horizontal planes normal to gravitational direction, and inclined layers reflecting lateral accumulation from currents.

In a normal sequence, the oldest rocks are at the bottom and the youngest at the top (law of superposition). Every area of deposition is related to an area of denudation the two being linked by an area of transportation. So, one is being reduced while the other is building up.

Each sedimentary rock reflects the physical geography of its time of formation.

Sedimentation is essentially a sorting of disaggregated material, according to density and solubility.

2 Sedimentary facies

If conditions of deposition were uniform, one thick, unstratified, uniform sedimentary unit would form. When conditions of sedimentation change, e.g. rates of sedimentation, type of sediment supplied, climatic change, then the lithology will change. Faunal changes are likely to accompany these changes, and the result overall is facies change and stratification. This is the pattern resulting from "positive" movements of the earth's crust.

Alternatively, "negative" movements may stop the sediment supply. This will result in erosion of the top stratum. When sedimentation is returned there is likely to be a marked lithological change and an unconformable relationship between the two strata.

Repetition of strata may also occur, and is known as cyclic sedimentation. These frequently consist of alternating sandstones, limestones, shales, coals, seat earths, etc. The conditions at the site of deposition must therefore have changed radically between marine, deltaic, brackish, lagoonal and continental environments. This implies changes in sea level, either relative or actual, and the sedimentary sequence is due to the movement.

Not all rhythmic sedimentation is due to sea level change. Climatic (seasonal) fluctuations lead to alternating clay and silt bands producing varving. This reflects seasonal variations in rainfall and erosive/transportational power. This form of varve deposition is generally associated with glacial regions. Coarse material within the ice is released during the phase of melting in summer. This leads to the paler, coarser silt bands in the lake deposit. In winter, melting is at a minimum and only fine material is supplied to the lake giving the thin clay layers.

3 Diagenesis

Diagenesis is a difficult process to define, as the points at which the process begins and ends are difficult to determine.

Diagenesis includes all the processes which turn wet sediments into relatively stable rock under pressure and temperature conditions not far removed from those acting at the earth's surface. These processes begin at the time of sedimentation with the deposition of the sediment. The process continues after deposition and tends towards mechanisms akin to (burial) metamorphism.

Perhaps the main factors affecting the process of diagenesis are porosity and permeability. These allow movement of solutions within the sediment. Chemical reactions take place within pore spaces — i.e. are the "crucibles where reactions occur between pore fluids and components of the rock".

The reactions vary, as rarely does one get homogeneous sediment. They are usually mixtures of:

1. Sandy material
2. Muddy material
3. Carbonate material

So we have rocks which contain:

(a) Free silica — quartz, opal, chalcedony etc.
(b) Carbonates — calcite, aragonite, dolomite etc.
(c) Clay minerals — illite, kaolinite, montmorillonite etc.

Auxiliary minerals, e.g. ferromagnesians, feldspars, micas, occur in lesser quantities. So, in general, minerals rare in igneous rocks remain rare in sedimentaries also.

The main phase of diagenetic change, however, takes place much later, and is probably related to the occurrence of the first set of joints which are non-tectonic. These provide channels for escaping gases, water and dissolved salts, so changing the chemical equilibrium of the system.

4 Diagenetic effects

1. Pure Clay

Clays are the decomposed and recrystallized products of unstable minerals, and these tend to adsorb metal ions (Ca or Mg in montmorillonites, for example) onto their surfaces. This addition to the lattice binds the montmorillonite flakes and turns the sediment into a mudrock.

The same minerals may undergo base exchange and show an increased K content. As soon as joints appear, water leaves the sediment due to the weight of overburden. A decrease in volume by, on average 50% takes place, and flaky minerals rotate from their original random orientation, to a position parallel to the bedding. In this way, a mud becomes laminated.

2. Pure quartz sand

This material resists compaction and hence there is little change in volume. Solution and cementation may be maximized if carbonates are introduced in fluids; attack on the quartz grains often results in a finely etched surface. If the solution is a silicate one, it is likely that quartz will grow in optical continuity about the quartz grains and the original clastic texture will be lost. The rock would be a quartzite.

3. Pure Limestones

Calcareous muds or carbonate oozes are both essentially impermeable until the first joints appear. Very pure carbonates recrystallize rapidly. With the introduction of magnesium from sea water, dolomite crystals begin to appear, and with the development of joints and fissures, dolomitization may increase to envelop the whole rock.

2.3 Sandstones

Sandstones are clastic rocks in which the majority of the grains fall within the size range 0.06 mm to 2 mm. This covers all particles whether they are formed of quartz, ferromagnesian minerals, shelly material or rock particles. The pore spaces are either wholly or partly filled with grains of < 0.06 mm size, or a cement, or both.

Cementing materials vary in composition, although the most common types are:
(a) Silica
(b) Calcite
(c) Iron oxide
(d) Clay minerals

Sandstones are divided into 2 broad categories:
(a) Clean sandstones which have a cement and no matrix.
(b) Dirty sandstones which have a matrix and may or may not have a cement.

Each type of sandstone tends to be associated with certain rock suites.

1 Distinguishing features of sandstones

Three factors distinguish a sandstone:
(a) Source area.
(b) Length and intensity of weathering in the source area and the duration of transport.
(c) Manner in which clastic grains are laid down in the basin of deposition.

The source area is composed of surface rocks, including volcanic rocks, pre-existing sedimentary rocks, etc., and also rocks with a sub-surface origin, e.g. intrusive igneous which may be exposed at the surface at that time. The initial sediment from these rocks will depend on the weathering process in the first instance, e.g. rock exposed to prolonged chemical weathering yields grains of quartz, chert and muscovite. Other materials are dissolved or broken down and converted to clay minerals. Mechanical weathering results in the accumulation of fragments of pre-existing rock types plus quartz, chert and muscovite. There are, therefore, two types of grains:
(a) mature e.g. stable grains of quartz, chert and muscovite
(b) immature

The chemical weathering of old crystalline rocks will yield quartz and muscovite, whereas mechanical weathering would additionally yield feldspars and ferromagnesian minerals. Examination of sediments with regard to their mineral content will give information of weathering processes and the source of the material. The shape of the grains tells us less as these are often inherited, e.g. a slate will always provide flaky fragments, and a granite, equant grains.

The degree of angularity or roundness is a measure of the intensity and the duration of transportation. Textures are largely influenced by the mechanism of sedimentation.

2 Classification of sandstones

There are many ways of classifying sandstones, e.g. grain size; but a genetic one should be aimed at, i.e. one related to origins where groups are linked by a common origin. The best judge of a classification is application in practice, and three characteristics in particular are useful:

(a) Source area (provenance)
(b) Maturity of the grains
(c) Fluidity of the depositing medium

Each characteristic provides an index.

The provenance is established from the nature of the grains: if stable minerals are present, then a surface origin is usually indicated. If the rock has immature grains, it is of magmatic origin. The ratio between fragments of immature grains and silica will provide a provenance index.

The maturity factor is then used to provide another index. The stable mineral which is concentrated from a weathered and eroded igneous rock is quartz and therefore the ratio of stable to unstable grains provides a check on the maturity index.

Effectiveness of sorting depends on the fluidity of the depositing medium. Dirty sandstones are deposited from turbidity and density currents, while clean sandstones come from dilute suspensions. The index is the ratio of sand size material to matrix material. The degree of particle roundness is also a feature of maturity.

3 Clean sandstones

This category is commonly associated with limestones, conglomerates and dolomites. This is a characteristic succession of stable shelf sea areas. The rock suite tends to be characterized by abundant fossils and to develop in relatively thin rock units.

Such rock types as clean sandstones characteristically display diagnostic sedimentary structures, e.g. cross-bedding, ripple marks. Bottom structures tend to be rare.

Clean sandstones are usually well sorted and are formed predominantly of quartz, plus other stable and resistant minerals. The grains are normally well rounded and the cementing agent is usually, though not always, calcite. The cementing material is introduced after the deposition of the grains.

4 Dirty sandstones

These are associated with areas of active tectonic processes, rapid uplift and intense erosion. They are characteristic of destructive plate margins, and may be found on land or in the sea especially marine trenches and are often associated with interbedded shale and chert. They tend to be extremely thick. Greywackes are the most common marine trench deposits, and commonly have abundant bottom structures, e.g. castes, grooves, slump structures, etc. indicating deposition from strong currents called density or turbidity current. Sedimentary structures, like cross bedding are rare. Greywackes tend to be poorly sorted. The most common sedimentary structure is graded bedding with a fine-grained component throughout. Greywackes are composed virtually of anything, including unstable minerals which would normally be destroyed by longer weathering or transportation histories.

The clastic material is in a matrix of mud, fine mica, chlorite, etc. commonly derived from low grade metamorphic rocks. The amount of matrix largely determines whether or not a rock is a greywacke.

5 Feldspathic sandstones

These sandstones contain feldspar grains, with quartz, mica and rock fragments.

Feldspar decays readily and cannot stand a long transport history unless it is in density current, i.e. very fast.

Often feldspathic sandstones have angular grains and ferruginous cements. An igneous provenance is often indicated. The iron oxide cement indicates sub-aerial oxidizing conditions. Mechanical breakdown of an igneous rock occurs in a sub-aerial, sub-arid or arid environment, in which the detrital material is rapidly buried and therefore escapes chemical weathering. If the feldspar grains show considerable decomposition within these sandstones, the rock becomes relatively weak and friable.

2.4 Mudrocks

Mudrocks are classified as clastic rocks which have greater than 50% of their particles smaller than sand size (0.06 mm). The bases of the classification are whether or not the rocks have a cement, whether or not they are laminated, and their grain size (Lundegard and Samuels, 1980).

Lamination in mudrocks consists of layers up to 10 mm thick, and thinly laminated usually refers to layers up to 2.0 mm. Lamination may occur as a result of grain size variations, fabric, most commonly the orientation of platey mineral grains, or colour.

The way in which these criteria are employed in identifying mudrocks is shown in Fig. 2.10.

Cementing agents in mudrocks are various, but most commonly consists of calcium carbonate. The cementing material is frequently unequally distributed throughout the rock, so that variable rock quality and/or the formation of nodules tend to be quite common.

2.5 Carbonate Rocks

These represent a maximum of 15% of sedimentary rocks in the geological column, and are therefore of less importance quantitatively than sandstones (33%) and mudrock (50%).

The mineralogy is simple, being principally of calcite and dolomite. Aragonite is the less stable form of calcite. Siderite (Fe CO_3) is also present.

The unconsolidated sediment is largely of unstable aragonite (mainly shell fragments) whereas the ancient carbonate rocks consist of calcite and aragonite. This transformation must take place in the solid form. This is clear from the primary structures and organic remains. The shells are preserved in outline, but are overprinted with a new granular fabric. The aragonite of the original shells is recrystallized and replaced by calcite. The nature of this process is not fully understood, although it has been carried out experimentally by immersing aragonite in distilled water at 23 °C for 100 days.

1 Terminology and classification of carbonate rocks

Carbonate rocks are difficult to classify. There are several approaches based upon field relations, structures, textures, minerals. But there are as many classifications as specimens.

Any terminology for modern carbonates is not applicable to ancient carbonates because of their sensitivity to changes with temperature and pressure.

The best classification combines both genetic and descriptive qualities. Information is required on origins, composition and texture. Most classifications include the distinction between calcite and dolomite, e.g. Fig. 2.4,

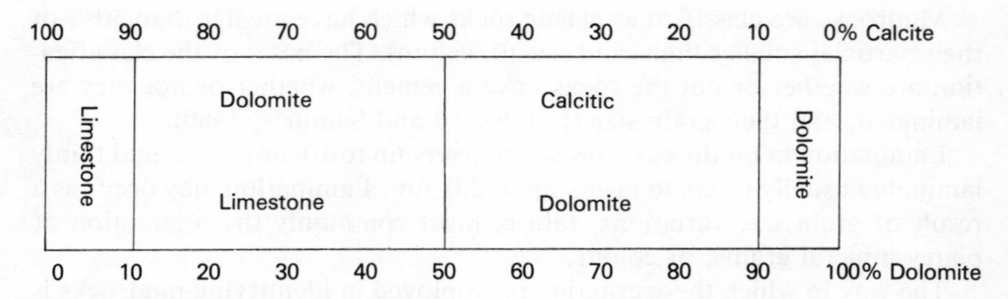

Fig. 2.4 Classification of carbonate sedimentary rocks.

then identify the types of grain which make up each limestone type. Limestones can also be classified in terms of their original organic materials.

This is applicable for all carbonate rocks, each subdivision being based on the fundamental classification. In clastic carbonate rocks, further subdivisions can be arrived at on the basis of grain size.

In terms of engineering requirements, this classification is probably most useful in explaining variations in the solubility of limestones, dolomite (the mineral, $CaMg(CO_3)_2$) being less soluble than calcite ($CaCO_3$). However, in terms of identification, differentiating a dolomitic limestone from a calcitic dolomite may not be very straightforward.

The main competence of carbonate rocks tends to accrue from recrystallization. This involves a process called pressure solution which will operate in these rock types under overburden pressures. This is not generally the case in silicate rocks except at locations of concentrated stress like grain boundaries, and pressure solution in silicate rocks is often an indication that they have been extremely strained. Even in carbonate rocks however, crystallization is not uniform. The quality of the re-crystallization varies in rocks of the same age, and same uniform calcite composition. The original fossils forming the rocks are different however, and this may well be the cause of the variation. In bioclastic limestones, that is ones formed from fragmentary shell remains, the rock, with sufficient application of time and pressure, is likely to appear uniformly crystalline. Very often, organic remains can only be identified in thin section. In limestones which are formed from whole or substantial parts of animals — whole shells, colonial corals or sea lilies (crinoids), for example, then any original in-built anisotropy will tend to be preserved for a long period. This preferred weakness may be across the top surfaces of the shells, which tend to have the same orientation anyway, or down the outer surfaces of individual corals or crinoids. The fossil boundaries continue to act as barriers to crystal growth across them, the process which would ultimately destroy the source of the anisotropy.

In rock samples used for strength testing, very large differences in strength can occur depending on the direction of loading with respect to the source of anisotropy. Limestones are also frequently quarried as construction materials. When crushed for aggregates, the shape of the aggregate can vary considerably because of the preferential lines of breakage of the rock. It should be emphasized that several of these changes can occur within a single quarry face, the variation reflecting small changes in environment in shallow warm seas tens or even hundreds of millions of years ago.

2.6 Metamorphic Rocks

Metamorphic rocks are formed by alteration of parent rock through heat, pressure and chemical action of fluids and gases. The parent rock may be either sedimentary, igneous or some type of metamorphic rock itself. Three types of metamorphic rocks may be recognized (Fig. 2.5):

1. Regional: This involves the large scale action of heat and pressure producing a wide range of new minerals and a series of crystalline rocks with a distinctive fabric resulting from mineral re-orientation e.g. slates, schists, gneisses.
2. Contact: Involves heat almost exclusively and is normally associated with igneous intrusions. It produces suites of "flinty" homogeneous often characterless rocks called hornfelses, and often includes marble. It is also referred to as thermal metamorphism.
3. Dynamic/: Burial This involves intense localized stresses tending to cause dislocation of minerals and break-up of rocks.

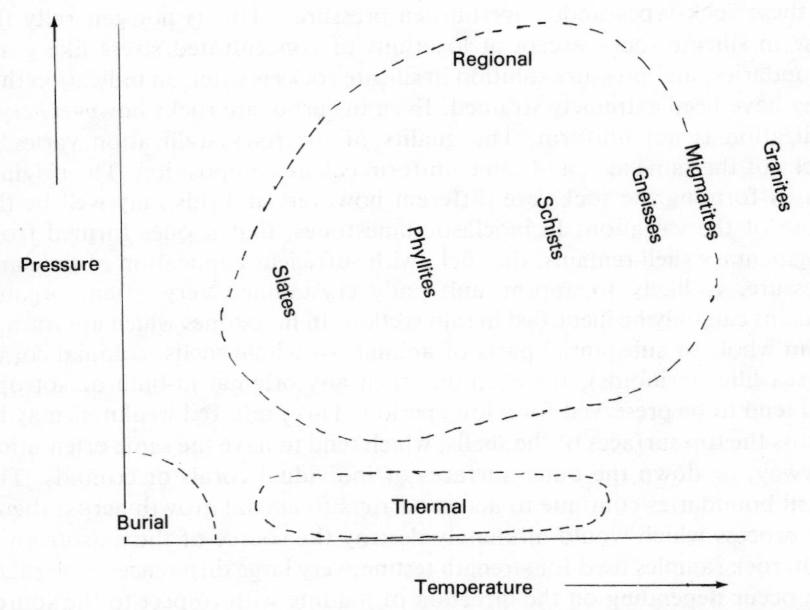

Fig. 2.5 Classification of metamorphic rocks based on temperature and pressure regimes.

It is clear from Fig. 2.5 that metamorphism is often found as a progressive series of changes depending largely on distance from the source of the heat and/or pressure. The severity of the metamorphism is marked by both mineral and textural changes, although the precise lines of the different zones of metamorphism are defined by the first occurrence of a particular mineral. The extent of metamorphic change is referred to as metamorphic grade, high grade rocks being the ones which have undergone most change.

The effects of metamorphism include:

(1) Chemical recombination and growth of new minerals, with or without the addition of new elements from circulating fluid and gases
(2) Deformation and rotation of the constituent mineral grains
(3) Recrystallization of minerals into larger grains

The net result is a rock of greater crystallinity and hardness, often possessing new structural features which result from the re-orientation of grains. Thus metamorphic rocks develop textures that are unique to these rocks and which help in the identification of such rocks.

The major textural difference in the various metamorphic rocks is in the "orientation" or "alignment" of the crystals. The other textural difference lies in the size of the crystals. Based on the alignment of crystals, there are two general textural groups — "foliated" and "nonfoliated" textures. In a well foliated rock, platy or leaf-like minerals such as mica or chlorite are nearly all aligned parallel to one another so that the rock splits readily along the well-oriented, nearly parallel cleavages of its constituent mineral particles. On the other hand, the nonfoliated rocks are composed of randomly oriented minerals, so that the rock breaks into angular particles.

Metamorphic rocks usually occur in zones based on the progressive changes in mineral assemblages found away from the source of metamorphism. The changes depend on initial composition of the rock and the environment. Each mineral assemblage is characteristic of a specific range of temperature and pressure. Many of the minerals of sedimentary and igneous rocks dominate the metamorphic rocks also. These include quartz, feldspars, calcite, micas, amphiboles and pyroxene. But the processes of metamorphism also create minerals which are uncommon in the sedimentary and igneous rocks, especially garnet, epidote and chlorite.

The relationship between texture and mineralogy is important for the identification of metamorphic rocks. The texture is by far the most important attribute of these rocks for civil engineers as it determines the physical properties of the rock in most cases, the foliated metamorphic rocks being highly anisotropic. Slaty rocks are fine-grained and formed mainly by the reorientation with slight growth and bonding of dominantly platy minerals. As the metamorphic grade increases, the minerals grow and become flexured themselves, disturbing the rigid, highly planar slaty cleavage. Thus, although schistosity is a clear plane of weakness, it is often wavy.

Up to this stage, the mineralogy of the rocks is dominated by minerals of the mica group, platy in form, and easily aligned. At temperatures and pressures at the higher end of the schistose rocks, new massive minerals make an appearance, usually garnets, which begin to disturb the "flow" of the schistose partings. The garnets force the micas to deflect around them, and the more garnets there are, the less persistent is the schistosity. This break-up of schistose fabrics also occurs as platy minerals become significantly less important than massive ones. Feldspars and quartz appear in gneissose rocks with increasing abundance. The remaining micas are left to form thin bands separated from each other by thicker quartz-feldspar bands.

Only at very high metamorphic grades where some melting occurs are the bands significantly disturbed, and at this stage, solid chemical changes make way for magmatic changes to take over. It is clear though that if the mineralogy of gneisses is dominated by quartz, feldspar and mica, then on re-melting and crystallization, the most likely rock to be produced is a granitic one.

2.7 Classification of Rocks

1 The use of feldspars in rock classifications

Feldspars are useful because of their common occurrence and wide variety, and are particularly useful in classifications of igneous rocks.

Feldspars are the most important group of rock forming minerals, and detailed identification is mainly carried out by measurement of their extinction angles in thin section.

Igneous rocks

Igneous rock types are numerous and it is therefore difficult to find one totally adequate basis for classification. The main bases are mineralogical, and an assessment of the norm of a rock quickly indicates the important role of feldspars (Fig. 2.1). However, rocks similar chemically will, due to differences in crystallization and cooling laws, produce different minerals. This is directly related to Bowen's Reaction Series which shows the general order of crystallization from a magma (Fig. 2.2).

Very few igneous rocks contain no feldspars, and they are often the dominant mineral group present.

An initial classification can be outlined based on the varying proportions of alkali feldspar and plagioclase feldspar (excluding albite). These often occur together, though also occur separately in some rocks. So we can differentiate between dominance of:

(a) alkali feldspar,
(b) plagioclase feldspar,
(c) alkali and plagioclase feldspar in roughly equal proportions.

Metamorphic rocks

Their classification and nomenclature is largely based on feldspars. The greatest onset of feldspars in this context is as a geothermometer e.g. the change from an original sedimentary rock which is unlikely to have an important feldspar content, to a low grade metamorphite which is often similar to the original rock; to a medium grade metamorphite in which feldspars begin to appear in some quantities; to the high grade metamorphics which possess an important content. So, feldspars increase in abundance with increasing pressure and temperature conditions. This has to be modified in some cases, but holds good as a general rule. To illustrate this, consider the example of metamorphism of a shale.

Sedimentary	Low Grade metamorphic	Medium Grade metamorphic	High Grade metamorphic
SHALE clay	SLATE clay begins transformation to mica; mica too small to see, but imparts cleavage to the rock.	SCHIST mica grains larger giving clear foliation	GNEISS mica has transformed to feldspar giving the rock a banded or layered aspect.

Sedimentary rocks

Shales/clays are the most common sedimentary rocks as these are the main weathering products of feldspar. However, feldspars are unstable during weathering and play a minor role in classification. The usefulness of feldspar in classifying sedimentary rocks is two-fold.

(a) They will give a clear indication of the parent material which was weathered and eroded to form the sediment; and

(b) They will give some indication of the transport history. Feldspars cannot survive long transportation.

Arkose is classified quantitatively on its feldspar content. The content of 20–25% to make the rock an arkose may be as high as 35% depending upon the above two factors. An arkose is derived from the mechanical disintegration of a granite or gneissic rock, each of which contains a high feldspar content. The transport history needs to be short.

The derived grains of a clastic sediment depend to a large extent upon the relative immunity to destruction, usually by chemical means, displayed by the minerals. Hence, the destruction of an igneous rock, say a granite, may lead to the formation of a sand, made up to a large extent of quartz, with white mica, i.e. chemically, relatively inert substances, relative, that is, to feldspars, which are easily destroyed.

2.8 Identification of Rocks

The rocks, like minerals, can be identified using several methods many of which involve the use of specialized equipment and/or techniques. In the field one does not have access to such equipment and thus the identification of rocks in the field is based on simple properties determinable in hand specimens. The three characteristics that are used in the identification of all types of rocks — igneous, sedimentary, metamorphic — are

* texture
* mineral composition
* properties specific to certain rock types

The figures and tables used here are modified versions of those used by

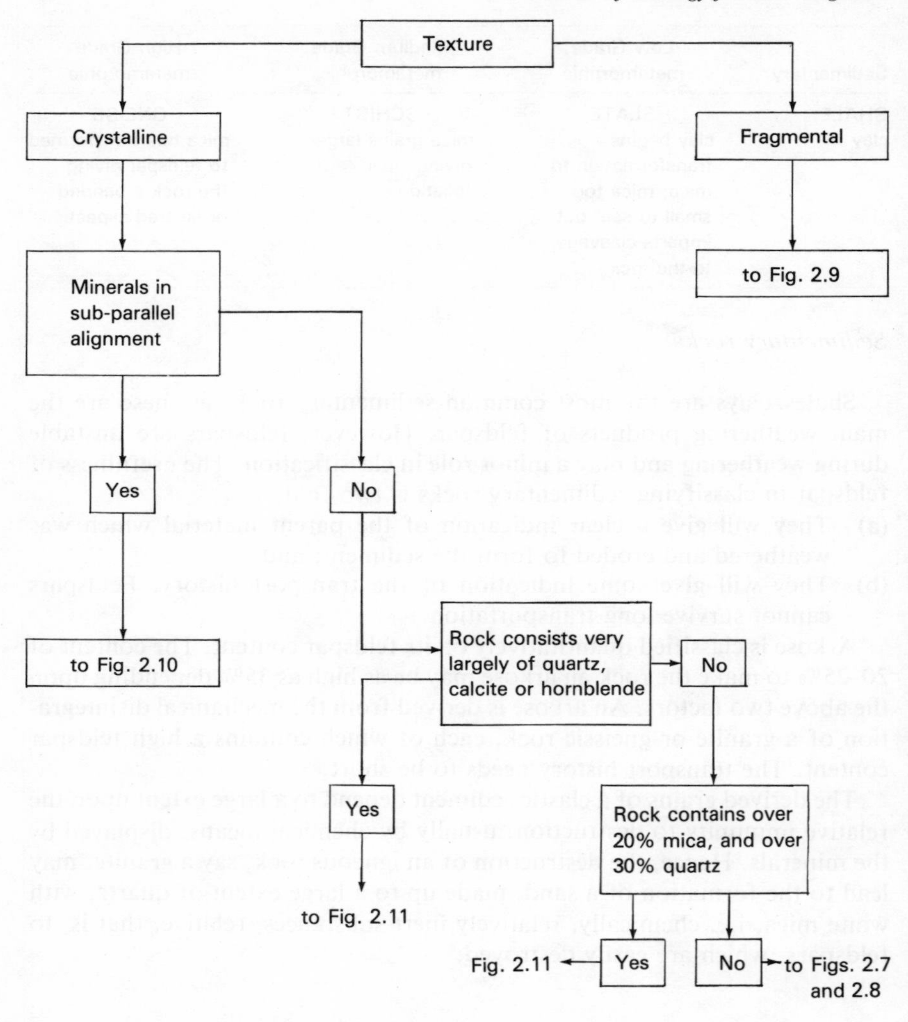

Fig. 2.6 Flow diagram for rock identification and classification.

the Open University in their Reference Handbook for the Study of Minerals and Rocks.

The general procedure of identifying the rocks is to first determine the major type of rock (i.e. whether igneous, sedimentary or metamorphic) through their diagnostic textures (Fig. 2.6). For instance most igneous rocks are composed of interlocking crystals. This is especially obvious in rocks containing large crystals. For fine-grained igneous rocks such interlocking crystals are visible only under the microscope and in hand specimens these rocks appear massive and homogeneous and in many cases may exhibit other diagnostic features such as porphyritic texture with some crystals large enough to be distinguishable; or gas holes called vesicles (as in vesicular basalt). When noncrystalline, the igneous rocks are glassy. Similarly most sedimentary rocks are identifiable through their clastic texture, or when crystalline through their monomineralic (also the type of minerals) composition. In most clastic sedimentary rocks the layering due to stratification is often quite evident. Most metamorphic rocks are distinguished from other types of rocks by their characteristic foliation. Non-foliated metamorphic rocks may also exhibit some elongate grains or other linear texture resulting from directional stresses. Some nonfoliated rocks are also monomineralic and minerals are coarse-grained and identifiable.

1 Identification of igneous rocks

The igneous rocks are identified mostly on the basis of their texture and mineral composition.

Texture in igneous rocks refers to size, shape and boundary relations between adjacent minerals.

Textures of igneous rocks develop primarily in response to composition and rate of cooling of the magma. Magmas located deep within the earth's crust cool very slowly. Individual crystals are more or less uniform in size and may grow to an inch or more in diameter. A magma extruded out upon the earth's surface, in contrast, cools rapidly and the crystals have only a short time to grow. Crystals from such a magma are typically so small that they can rarely be seen without the aid of a microscope, and the rock appears massive and structureless. If extremely rapid cooling took place, as would result if lava flowed into the sea or a lake, the lava would be quenched and the rock resulting from such a process would be a natural glass. An additional textural type may develop if the cooling history involves a period of more rapid cooling. Two distinct crystal sizes would probably develop; the large crystals, called phenocrysts, develop during the slower cooling period and are surrounded by smaller crystals which form during the period of rapid cooling. The latter determine the classification of the rock (Fig. 2.7).

Thus the major types of textures found in igneous rocks are:

(i) *Granular texture* (Fig. 2.7) composed of crystals that are large enough to be seen and identified without the aid of lens or microscope. Even though the average size of individual grains may vary from about 0.5 mm to several centimetres, the common granular rocks such as granite have grains averaging from 3 to 5 mm in size. Granular textures develop from magmas which cool slowly and commonly develop in intrusive igneous bodies.

(ii) *Medium texture* in which individual minerals can be seen, but are too small to identify with the unaided eye. These rocks mainly result from moderate rates of cooling found in small intrusive bodies like dykes and sills.

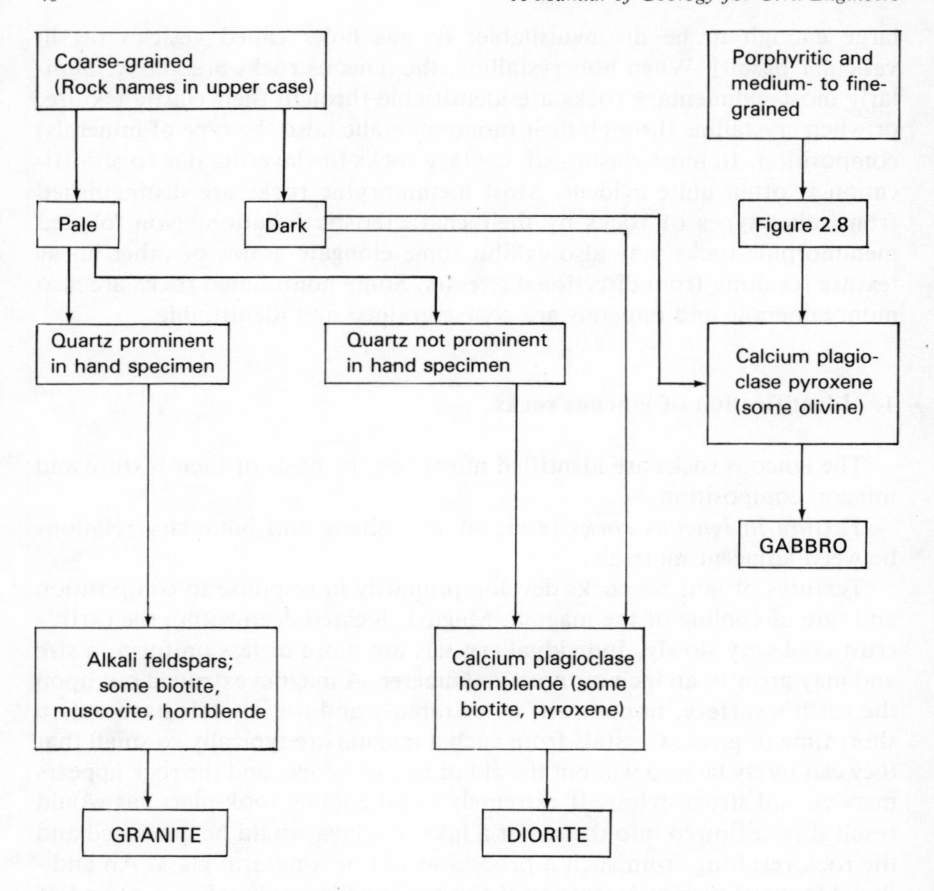

Fig. 2.7 Flow diagram for the identification and classification of plutonic igneous rocks.

(iii) *Fine texture* (Fig. 2.8) in which individual crystals are so small they cannot be detected without the aid of a microscope. Rocks of this texture appear massive and structureless. Fine textures are the result of rapid cooling of magma as in volcanic rocks, e.g. lava flows. Vesicles (spherical holes formed by gas bubbles as the rock cools) form near the tops of many lava flows.

(iv) *Glassy texture:* This texture is similar to that of ordinary glass. A glassy texture does not contain crystals even when viewed under high magnification. The dense glass has bright vitreous luster, conchoidal fracture with razor-sharp edges, and is transparent along thin edges.

(v) *Fragmental or pyroclastic texture* consist of broken fragments of ejected igneous material ranging from large blocks to fine dust. The rock may contain fragments of the wall rock surrounding the (erupting) vent, but it is composed mostly of fragments of ash, pumice and fine-

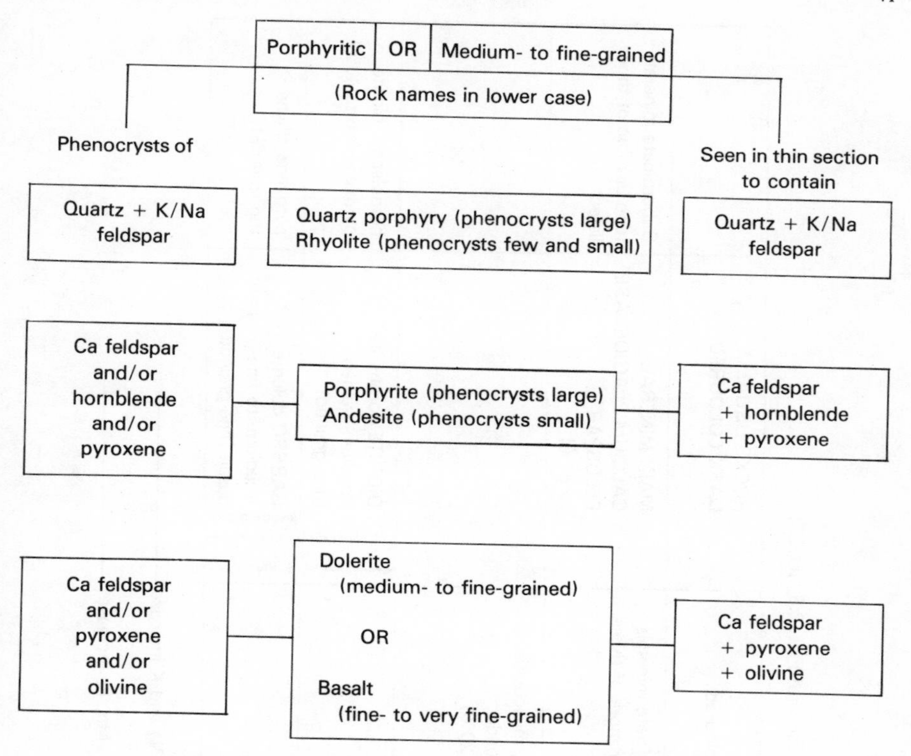

Note: In porphyritic rocks, it is an almost invariable rule that phenocryst minerals are also present in the ground mass. This can be confirmed in thin section when subordinate minerals can be identified. Remember however, that it is the size of the crystals in the ground mass and not the phenocrysts which determine the classification.

Fig. 2.8 Flow diagram for the identification and classification of small scale intrusive and extrusive igneous rocks.

grained rock. Pyroclastic rocks are the products of volcanic explosions or of pyroclastic flows, e.g. as at Mt. St. Helens.

(vi) *Porphyritic texture* (Fig. 2.8) i.e. composed of two widely different sizes of minerals — larger "phenocrysts" embedded in a finer "groundmass" — giving a spotted appearance. Porphyritic texture indicates two stages of crystallization — one for phenocrysts (relatively slower cooling and other conditions favouring the growth of larger crystals) and the other for the groundmass (faster cooling and other conditions favouring the growth of smaller crystals or glass).

Composition of Igneous Rocks (Table 2.4). Feldspars, quartz, mica, amphiboles, pyroxenes and olivine constitute over 95% of the volume of all common igneous rocks and are therefore of paramount importance in their identification. Identification of individual minerals in igneous rocks may pose some problems to the beginner. Of course if the rock is very fine-

Table 2.4 (for use with Figs. 2.8 and 2.7).

	ROCKS USUALLY PALE COLOURED (Some finer-grained rocks can be dark coloured e.g. rhyolite)		ROCKS USUALLY DARK COLOURED	
	MAIN MINERAL ALKALI FELDSPAR (> 50%)	Subordinate minerals (one or more of those listed)	MAIN MINERAL CALCIUM PLAGIOCLASE FELDSPAR (> 50%)	Subordinate minerals (one or more of those listed)
QUARTZ prominent in hand specimen (up to 30%), except in very fine-grained rocks	GRANITE quartz porphyry (phenocrysts prominent) rhyolite (phenocrysts not prominent)	biotite, muscovite, hornblende (up to 20%)		
QUARTZ not prominent in hand specimen, may be absent altogether, especially in basic rocks			DIORITE porphyrite (medium-grained) (fine-grained)	hornblende, biotite, pyroxene (up to 40%)
			GABBRO dolerite (medium- to fine-grained) basalt (fine-grained)	pyroxene olivine (up to 40%)

Arrows within table: Si increases ↑ — Na, K, Ca, Mg, Fe increase ↓ (alkali feldspar rocks); Si, Na, K increase ↑ — Ca, Mg, Fe increase ↓ BASIC ROCKS (calcium plagioclase rocks).

← Si, Na and K increase

Ca, Mg, Fe increase →

grained then the minerals in the hand specimen can only be identified through indirect means such as colour (generally, rocks rich in mafic minerals are dark coloured and rocks rich in felsic minerals are light coloured); presence of phenocrysts and mineral associations. Starting with the dominant mineral(s) try to determine the properties of all minerals in the rock — individually. You may not be able to determine all the properties for all the minerals, therefore pay extra attention to the diagnostic properties (e.g. look for feldspar cleavage, conchoidal fracture in quartz, perfect cleavage of biotite etc.). Mineral association should also help, for instance if abundant quartz is present, the rock is unlikely to be a gabbro. Write down the properties and names of observed minerals in the appropriate columns of your worksheets. Also make note of the relative abundance of the various minerals. Make note of other characteristics such as colour, presence of vesicles, etc. Note not only the intensity of the colour i.e. dark, medium, or light — but also the actual colour such as pink, gray, black etc. After you have recorded all the characteristics, refer to the tables of classification of igneous rocks. Just like you did for minerals, match the characteristics as you observed them against those listed in these tables and name the rocks.

2 Identification of sedimentary rocks

The principle of identifying the sedimentary rocks is similar to that of igneous rock, i.e. you examine the texture, mineral composition and other specific characteristics. Characteristics of sedimentary rocks have already been outlined.

Texture of sedimentary rocks: First determine whether the rock has a clastic or nonclastic texture. All clastic sedimentary rocks are composed of particles of broken rock transported, deposited, compacted and usually cemented to form a coherent mass. Nonclastic crystalline textures consist of a network of interlocking crystals. Such textures are similar to those found in igneous rocks but generally consist of one dominant mineral. The interlocking crystals are usually about the same size, and interlock to form a dense rock.

These textures are further subdivided as outlined below:

Clastic Textures (Fig. 2.9, Table 2.5): The basic criterion for classifying clastic textures is grain size, with subordinate subdivisions made on the basis of rounding, sorting, and cementation. Note that the degree of roundness depends upon the amount of abrasion the grains have been subjected to. In a general way, grain size and rounding are rough measures of distance over which the particles have been transported. Large angular boulders indicate a nearby source because any significant transport by streams would rapidly round off the corners and wear down the size.

Sorting refers to the range of various sizes of particles. It is a very important textural characteristic as it may provide clues concerning the history of

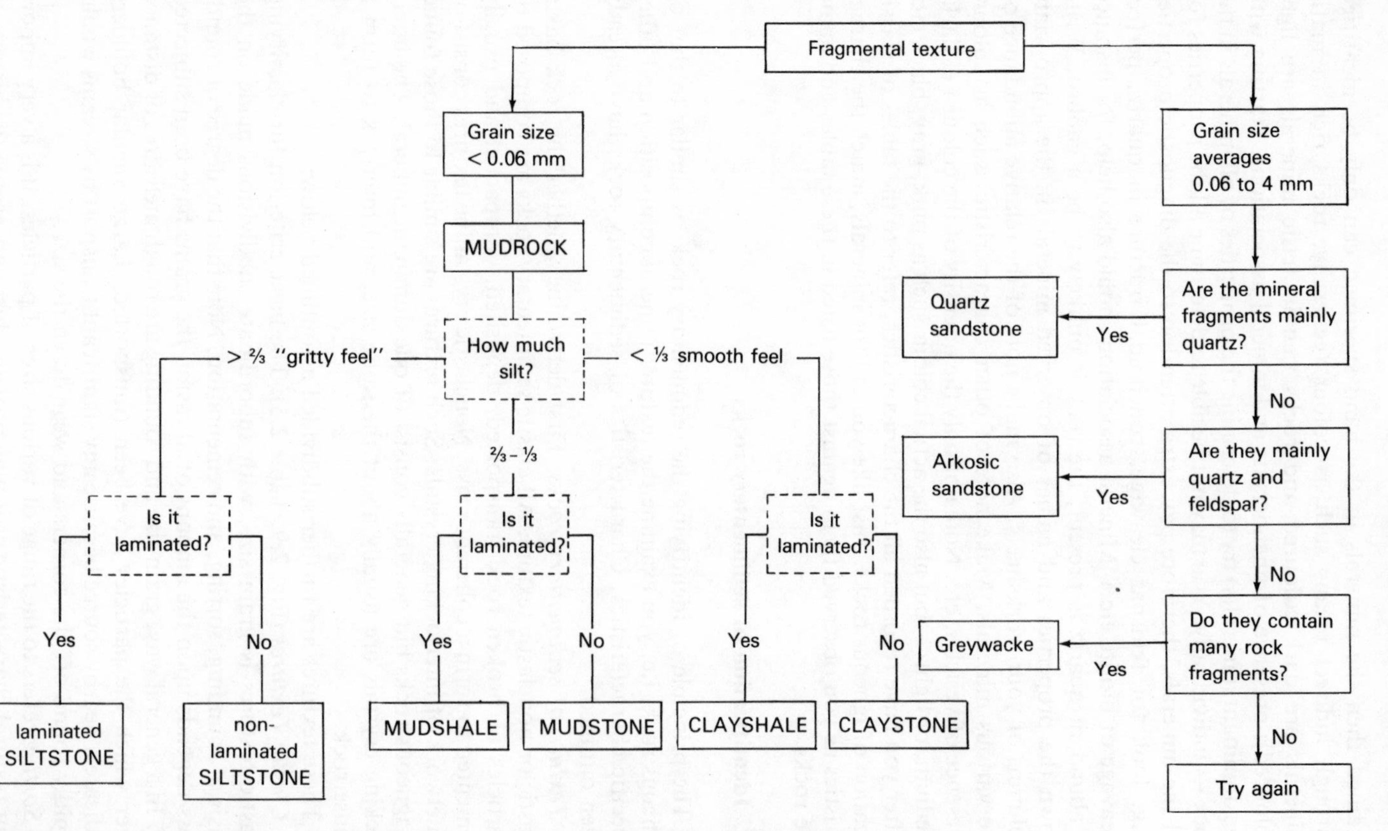

Fig. 2.9 Flow diagram for the identification and classification of clastic sedimentary rocks.

Table 2.5 (for use with Fig. 2.9).

Approximate average grain size	Quartz content 95%	Feldspar prominent, but quartz still dominant	Rock fragments prominent, with quartz, feldspar, and muddy matrix	Cement
4 mm	CONGLOMERATES (breccias if angular)			Any of these can also have an intergranular chemically precipitated cement which may be siliceous (SiO_2), making them very hard, ferruginous (Fe_2O_3, making them brown or red), or calcareous ($CaCO_3$, making them 'fizz' with acid). If there is a lot of cement, its nature may prefix the rock name, e.g. ferruginous sandstone.
0.06–4 mm	QUARTZ SANDSTONES	FELDSPATHIC (ARKOSIC) SANDSTONES	GREYWACKES	
0.06 mm	SILTSTONES (coarser), MUDSTONES (finer) or CLAYSTONES (fine); non-indurated and laminated equivalents			

transportation and the environment in which the sediment accumulated. Well-sorted material is composed of one dominant size and usually one type of material.

After a sediment is deposited, it will be subjected to various degrees of cementation. The degree to which a sediment is cemented is an important textural characteristic in clastic rocks.

(Nonclastic) Crystalline Texture: Crystalline textures are described as coarse (greater than 2 mm), medium (2 mm to 0.06 mm) and fine or microcrystalline (less than 0.06 mm).

Skeletal Textures: Calcium carbonate may also be removed from sea water by organisms to make their shells and other hard parts. When the organisms die the shell material will settle to the sea floor and may be concentrated as shell fragments on a beach or near a reef. The texture of the resulting rock is similar to a clastic texture but the material is unique in that it consists of the skeletal fragments of organisms. Such a texture is referred to as skeletal and is of fundamental importance in many limestones. Various other terms used for such textures are "fossiliferous", "fragmental", "organic" etc.

Composition of sedimentary rocks: In as much as sediment is derived from any pre-existing source rock, one might expect the composition of sedimentary rocks to be extremely variable and complex. This is indeed true if the sediments are deposited close to the source area; but if weathering and erosion are prolonged, sedimentary processes will concentrate materials similar in size, shape and composition in separate deposits. Most sedimentary rocks are thus composed of materials which are abundant in other rocks and are stable under surface temperature and pressure.

Most sedimentary rocks are composed of only four constituents; quartz, calcite, clay minerals, and rock fragments.

Quartz is the most abundant clastic mineral in sedimentary rocks, beçause it is one of the most abundant minerals in the earth's crust, is extremely hard (Moh's hardness = 7), resistant, and chemically stable (formed at the end of the Bowen's Reaction Series). Sedimentary processes decompose and disintegrate less stable minerals, for instance feldspar, and concentrate quartz as deposits of sand. Silica in solution or in particles of colloidal size is also a product of weathering of igneous rocks and is commonly precipitated as a cement in certain coarse-grained sediments.

For the identification of quartz in sedimentary rocks, you need to concentrate on diagnostic and determinable properties and also the features of mineral association. In most sandstones, for instance, quartz is the dominant mineral; siliceous cement is identified on the basis of its hardness. Other cementing minerals are not as hard, and have different properties.

Calcite is the major constituent of limestone and is the most common cementing material in sands and shales. Due to widely differing modes of formation of limestones, calcite in limestones also exhibits widely different characteristics. Even though you may not be able to determine all the properties of calcite, either in limestones or in clastic rocks as cements, the diagnostic property of calcite — effervescence with dilute hydrochloric acid — is easy to determine.

Due to impurities, calcite in limestone may have different colours. Dense black limestones, on first inspection, are often identified as basalts. Most limestones, however, are light coloured and contain fossils, and, all limestones react with dilute hydrochloric acid.

Clay minerals are mostly found in clay size range, and develop from the weathering of silicates, particularly the feldspars. The abundance of feldspar in the earth's crust, together with the fact that it readily decomposes under atmospheric conditions, accounts for the large amount of clay minerals in sedimentary rocks. Rocks composed dominantly of clay minerals are termed mudrocks. They may be identified and classified as being cemented or non-cemented, and laminated or non-laminated. By wetting the rock and rubbing with the thumb, a sticky residue on the thumb would indicate a non-cemented mudrock. Clear splitting along the lamination (closely spaced bedding) often develops only when the rock is weathered. In fresh rocks, you will need to find a clean surface perpendicular to the bedding, and may have to use a hand lens to identify a lamination. Besides mudrocks, phyllosilicates may also be present in coarser rocks but there they are not the dominating minerals. Because of the extremely fine grains of clay minerals, the individual particles cannot be tested in hand specimen but the clay minerals are easily identified through their powdery nature (when disintegrated) and earthy smell. One of the most common clay minerals is kaolinite, a white chalky substance.

Several minerals found as sedimentary rocks are formed through evaporation of land-locked masses of seawater or saline lakes. Most common evaporites include halite (NaCl), also known as rock salt, and gypsum ($CaSO_4.2H_2O$). Other evaporite minerals are also found, but they are important only locally. Both halite (salty taste) and gypsum (soft, scratchable by finger nails) are easily identifiable.

There are other minerals, besides those listed above, that are found in sedimentary rocks. Dolomite, $CaMg(CO_3)_2$ for instance is found in many areas, along with calcite. Feldspars and micas may be concentrated in some sandstones if prolonged chemical weathering is inhibited.

Sedimentary Structures: A number of structures are unique to sedimentary rocks and are important identifying features of this rock class. The most significant are various types of bedding, layering found in most sedimentary rocks which is produced by physical or chemical changes that occur during deposition. The resulting layers may range from a fraction of an inch to many feet in thickness. In as much as sediments may accumulate vertically and laterally two basic types of layers result (1) horizontal layers and (2) cross-bedding.

3 Identification of metamorphic rocks

The general scheme of identification of metamorphic rocks is also similar to the previous two types of rocks, i.e. description of texture and identification of minerals.

Common textures of metamorphic rocks

(a) *Slaty texture* (Fig. 2.10, Table 2.6): Very fine foliation, producing almost rigidly parallel planes of easy splitting due to the nearly perfect parallelism of fine crystals of platy minerals, chiefly mica. As is evident the individual crystals/grains are not distinguishable.

(b) *Schistose texture:* Finely foliated, forming thin parallel bands along which the rock splits readily. Individual minerals/grains are coarse and are distinctly visible. Most minerals may be platy — or rodlike — chiefly mica, chlorite or amphibole, and equidimensional minerals — like feldspar, garnet, etc. may be present in lesser amounts. The foliation surfaces in schistose texture are wrinkled or bent.

(c) *Gneissose texture:* Consists of coarse-grained minerals that are weakly foliated. Most gneisses have a streaky, roughly banded appearance,

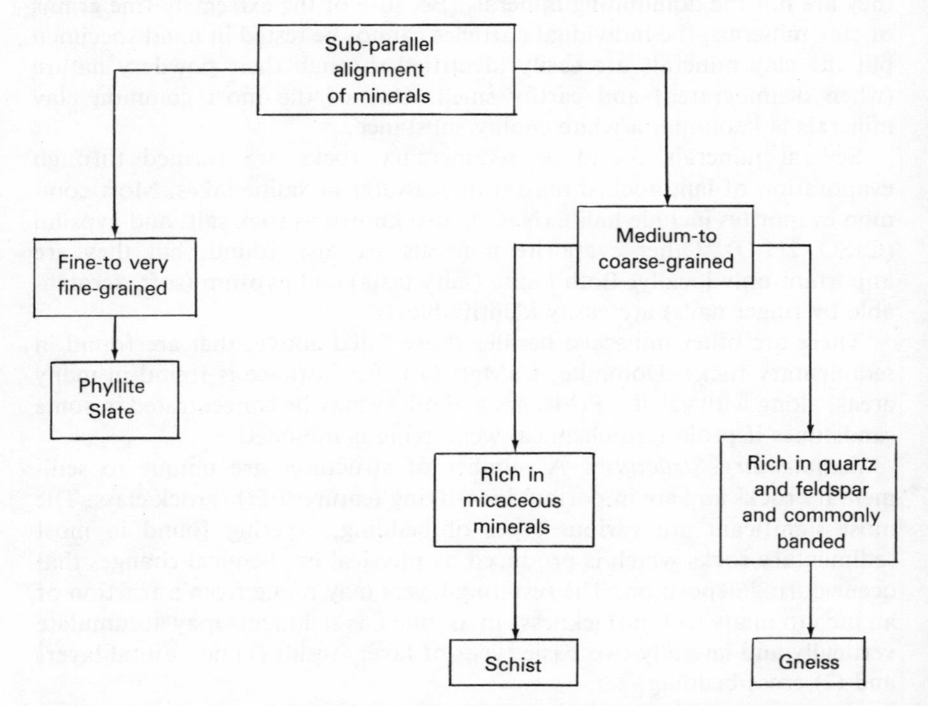

Fig. 2.10 Flow diagram for the identification and classification of foliated metamorphic rocks.

Table 2.6 (for use with Figs. 2.10 and 2.11).

	TEXTURE	COLOUR	MINERAL CONTENT	ROCK NAME	ORIGINAL ROCK
Obvious sub-parallel alignment or banding of minerals	Very fine-grained	Generally dark grey, green or red	Fine- or very fine-grained mica flakes in parallel orientation, often visible in hand specimen. Also very fine-grained quartz, feldspar, etc., generally only identifiable in thin section.	SLATE	Usually mud-rocks or volcanic ash.
	Fine-grained	Generally medium to dark grey, with mica-ceous sheen.		PHYLLITE	
	Medium- to coarse-grained, well-developed mineral alignment	Variable, depending on nature of dominant minerals commonly with mica-ceous sheen. Usually "silvery", green, or nearly black.	Most obvious will be one of: biotite, muscovite, or chlorite. Quartz and feldspar also usually present, and sometimes other minerals (e.g. calcite garnet, hornblende)	SCHIST	Mudrocks, greywacke, volcanic rocks.
	Medium- to coarse-grained, usually banded.	Pale	Generally quartz and feldspars dominant, along with micas, hornblende and other minerals.	GNEISS	As above, also arkose, some plutonic rocks.
Rare mineral alignment or banding	Fine- to very fine-grained and often of "flinty" aspect. Crystal sizes tend to increase with meta-morphic grade.	Generally dark (for all of this).	Generally only visible in thin section. Granular quartz abundant, commonly with randomly oriented and often sieve-like micas. May sometimes have slight mineral alignment or banding.	HORNFELS	Usually mudrocks.
	Medium- to coarse-grained	Usually white, grey red	Quartz greatly pre-dominates; may often have some muscovite too.	QUARTZITE	Sandstone
		Usually white, grey	Calcite greatly pre-dominates.	MARBLE	Limestone

caused by alternating layers of differing mineralogic composition. Gneisses are more "banded" than "foliated". Rocks containing gneissic layering are characteristically coarse-grained and represent a higher grade of metamorphism in which the minerals are stretched, interlocked, and completely rearranged. Banding may result from alternate layering of light coloured minerals (feldspars, quartz) and dark coloured ferromagnesian minerals (mica, amphibole).

(e) *Non-foliated textures:* (Fig. 2.11, Table 2.6) Consist of unfoliated or very faintly foliated textures composed of mutually interlocking mineral grains. In the coarse-grained variety, the minerals are large enough to be easily identified without the microscope, and are chiefly equidimensional kinds such as feldspar, quartz, calcite, etc. The finer variety called "hornfelsic" has mostly microscopic sized mineral grains, with very few visible grains. The grains are mostly scattered and consist of minerals that are common in zones of contact metamorphism. In many nonfoliated metamorphic rocks some linear features, that resulted from directional stresses, may be evident. For instance, a metamorphosed conglomerate may show elongated pebbles which have been stretched in response to directional forces. Deformation of limestone (into marble) will produce stretches or streaks of organic debris, or mineral impurities.

Minerals in metamorphic rocks: Even without being able to identify many of the minerals present, you should be able to adequately identify all the major types of metamorphic rocks by their textures. Notice that the foliated metamorphic rocks are named after their dominating texture — slate, schist, gneiss, etc.

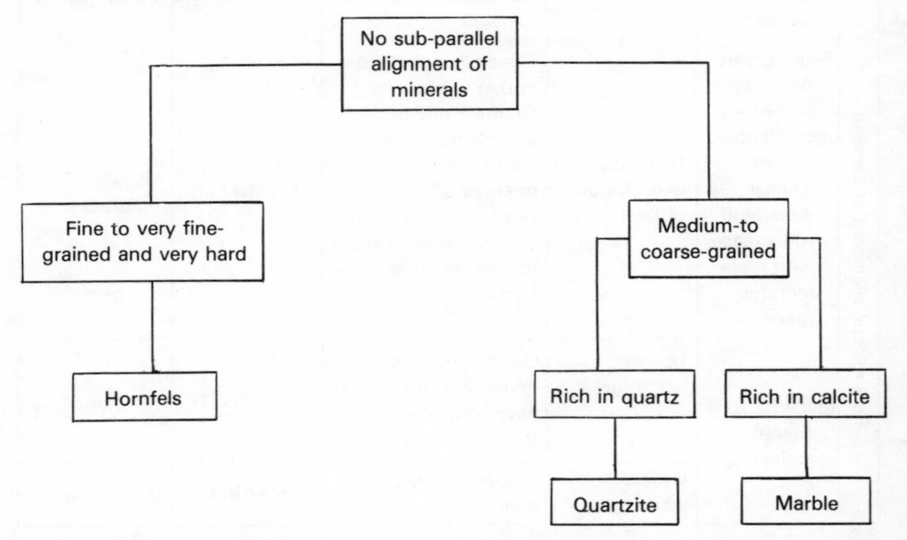

Fig. 2.11 Flow diagram for the identification and classification of non-foliated metamorphic rocks.

In order to arrive at the names of the metamorphic rocks, first determine the type of texture. You may already have a general colour and other specific features (e.g. coloured bands in marble). The special considerations for identifying minerals in metamorphic rocks are the same as in igneous rocks. Treat gneiss and schist as you would treat granite. Slates are so fine-grained that you can't identify the individual minerals. The two dominating minerals in slate are chlorite (green coloured) and muscovite (white mica). Most slates, however, have smaller amounts of other (accessory) minerals and they impart a distinctive colour to the rock. For instance some iron oxides will make the rock red while carbonaceous matter renders it black. You should have no problem identifying the mineral in a monomineralic unfoliated metamorphic rock such as marble (calcite) or quartzite (quartz).

2.9 Some Engineering Properties of Rocks

So far, in dealing with rocks, virtually no quantitative data have been introduced. In engineering, it is rather difficult to design on descriptions and some numerical values are required for use in design equations. The properties of rocks vary regionally and locally, necessitating a firm understanding of the different geological elements, their distribution, and variation. This is particularly important if results are to be extrapolated from laboratory tests. The design equations are from theoretical and applied mechanics. In almost all cases, some parameter representing a mechanical property of the rock must be inserted into the equation. The validity of the solution is no greater than the validity of the mechanical property utilized. Great difficulties have arisen in correlating the results of laboratory tests with those from in situ tests. Laboratory samples are prepared mainly from good sticks of core in the core box. What testing is carried out on the small core pieces, fragmented by extreme fracturing, by a lack of inherent strength, or sometimes just represented by a space in the box because the material was lost during sampling? In recent years, there has been a firm move towards field testing, and particularly the testing of irregular samples. Unprepared samples enable a great many tests to be carried out which extend not only the reliability of values chosen, but also the natural range of values present. In situ testing has also increased in importance, but these techniques are usually extremely expensive, and the final results are still not always representative when scaled-up for use in the full-scale design.

The properties of rocks of most importance in design are those of elasticity and strength. There are others, but these two only will be dealt with in detail.

2.10 Elastic Properties of Rocks

Laboratory tests are mainly carried out on cylinders of rock with length

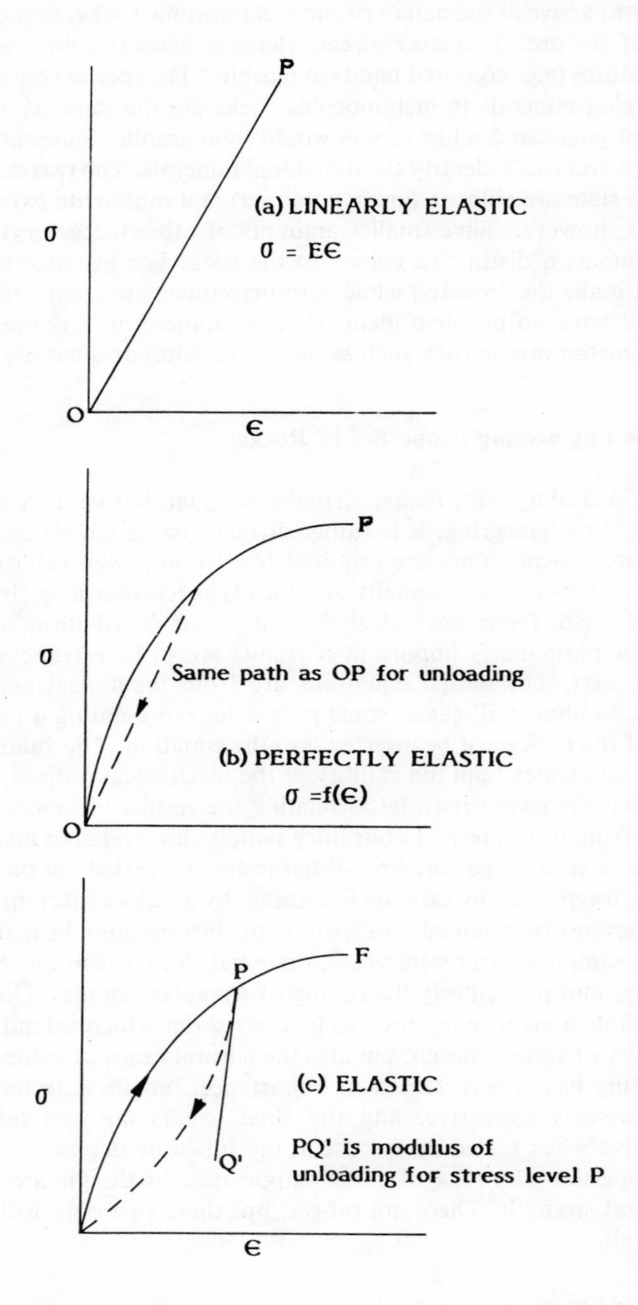

Fig. 2.12 (a) Linearly elastic material;
(b) Perfectly elastic material;
(c) Elastic material with hysteresis, showing loading and unloading cycle.

to diameter ratios of between two and three to one. Three conditions of elasticity are recognized to which rocks approximate.

1. Linearly elastic (Fig. 2.12a) where stress is equal to strain multiplied by Young's Modulus (E).
2. Perfectly elastic (Fig. 2.12b) where stress is a function of strain. The path of the test curve is the same for loading and unloading, although there is no unique modulus.
3. Elastic (Fig. 2.12c), where after unloading, strain returns to zero at zero stress even though it may be along a different path than that of the loading curve. This difference is called hysteresis.

1 Moduli of elasticity

In perfectly elastic and elastic rocks, there is no unique modulus.

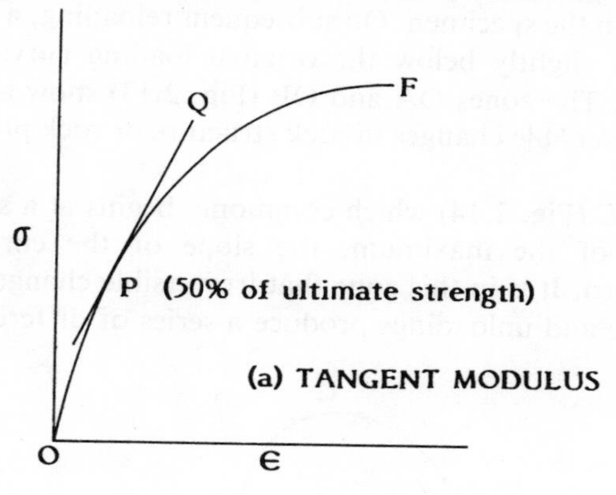

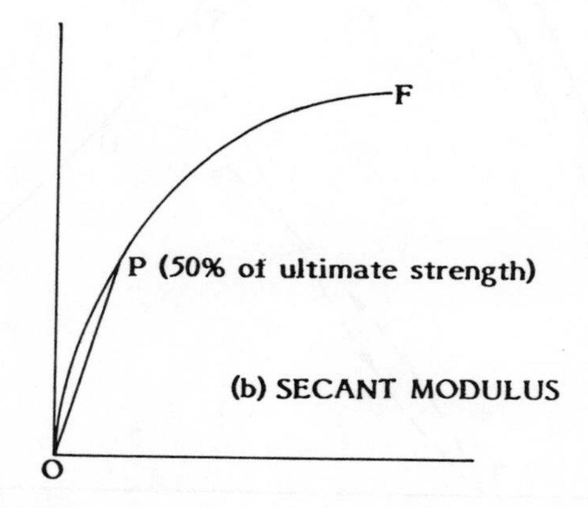

Fig. 2.13 (a) Determination of tangent modulus from stress-strain curve
(b) Determination of secant modulus from stress-strain curve

However, for any value of stress (σ) for example point P on Fig. 2.13(b), the slope of the tangent to the curve PQ is

$$\frac{d\sigma}{d\epsilon}$$

and is called the tangent modulus.

The slope of the secant OP on Fig. 2.13(b) which is $\frac{\sigma}{\epsilon}$ is called the secant modulus.

Tangent Young's Modulus (E_t) is measured at a stress level which is some fixed percentage of the ultimate strength, usually 50%. Secant Young's Modulus (E_s) is usually from zero stress to 50% of the ultimate strength.

In many rocks, unloading after the approximately straight line portion of the graph has been passed may lead to permanent set (ϵ_o) (Fig. 2.14), non-recoverable strain in the specimen. On subsequent reloading, a curve will be described which is slightly below the original loading curve, but which eventually joins it. The zones OA and OB (Fig. 2.13) show almost elastic behaviour and irreversible changes in rock structure or rock properties tend not to occur.

In the region BC (Fig. 2.14) which commonly begins at a stress level of about two-thirds of the maximum, the slope of the curve decreases progressively to zero. It is in this zone that irreversible changes occur, and successive loadings and unloadings produce a series of different curves.

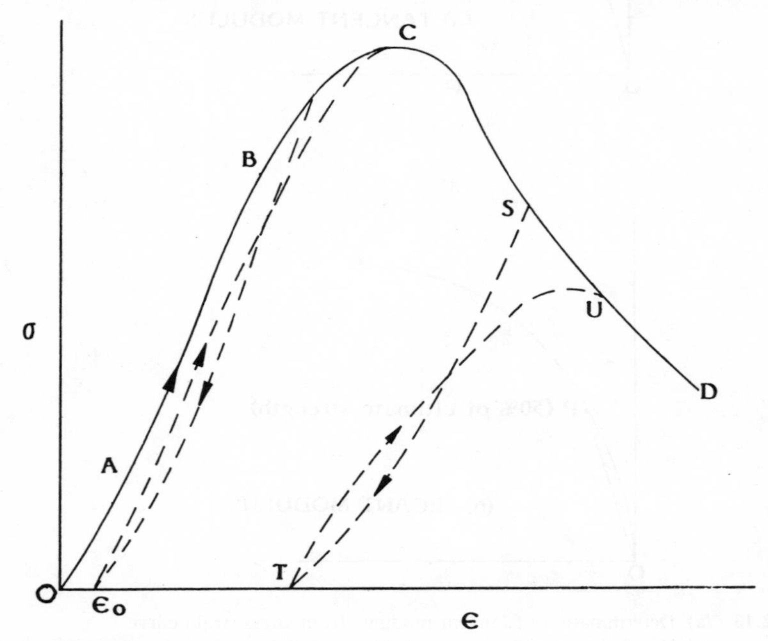

Fig. 2.14 The complete stress strain curve for rock.

The portion CD (Fig. 2.14) of the curve ranges from the maximum C and is characterized by a negative slope. Any unloading cycle at this region is likely to result in a large permanent set (T). The region CD is characterized by brittle behaviour, a fact generally obscured during laboratory testing which permits violent failure at around point C resulting in a total loss of cohesion across a plane. In underground rock systems, failure may begin at C and then progress steadily, the rock becoming less able to resist load with increasing deformation. However, the ability of partially failed rock within the range C to D to resist load, is extremely important.

2 Geological influences on the elastic behaviour of rocks

When dealing with real rocks, elastic theory applies only to greater or lesser degrees. The extent depends on the influences of:

1. Homogeneity, i.e. the physical uniformity of a body;
2. Isotropy, i.e. the directional properties of the material;
3. Continuity, i.e. joints, cracks or pore spaces which affect cohesion and therefore the transmission of an even stress distribution throughout the body.

Most rocks which approximate to linear elastic behaviour will be fine-grained, massive and compact. Volcanic and small scale intrusive rocks are examples. The linearity of the stress-strain relationship only lessens close to the point of failure, and this is referred to as "quasielastic" behaviour, (Fig. 2.15).

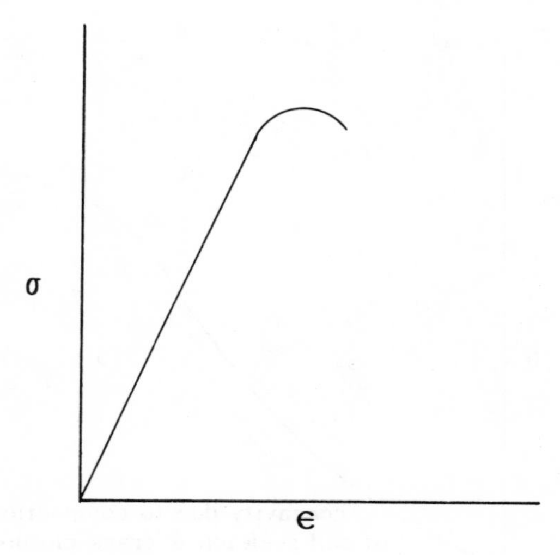

Quasi-Elastic Behaviour
e.g. Volcanic Rocks
Small-scale Intrusive Rocks

Fig. 2.15 Quasi-elastic behaviour in rock.

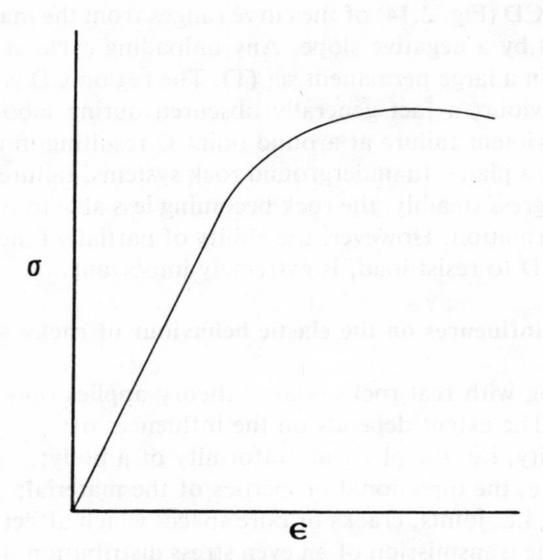

Fig. 2.16 Semi-elastic behaviour in rock.

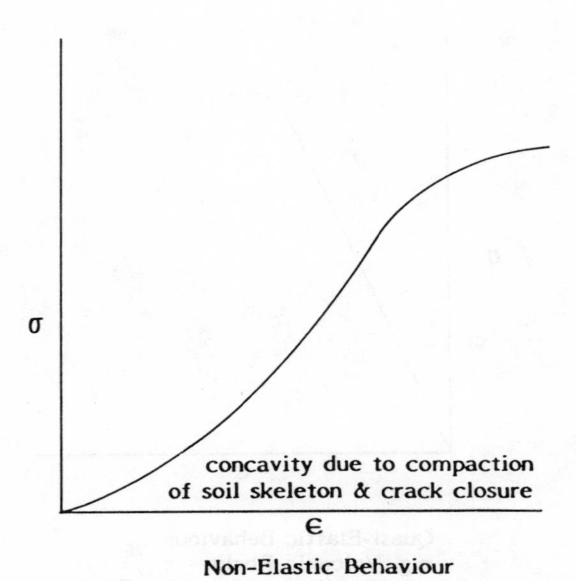

Fig. 2.17 Non-elastic behaviour in rock.

"Semi-elastic" behaviour (Fig. 2.16) is characteristic of plutonic rocks and fine-grained, compact sedimentaries of low porosity and moderate to high cohesion. The slope of the curve decreases with increasing stress. However, this type of curve is obtained from small laboratory specimens in which the effects of inhomogeneity and anisotropy are accentuated, possibly giving an exaggerated view of non-elastic behaviour. In the mass, such rock may be amenable to elastic analysis.

Rocks with low cohesion, and high porosity are "non-elastic" (Fig. 2.17) and are represented by many of the weaker clastic sedimentaries. They are not usually amenable to analysis based on elastic theory. The curve normally shows an initial stage of increasing gradient due most probably to compaction of the granular 'skeleton' and closure of cracks. A region of near linear deformation follows. Such rocks exhibit highly variable stress-strain characteristics.

3 Poisson's ratio

This is often a difficult parameter to measure. For elastic or approximately elastic rocks, a constant value of Poisson's ratio is often applied. This is acceptable for rocks with a high Young's Modulus value.

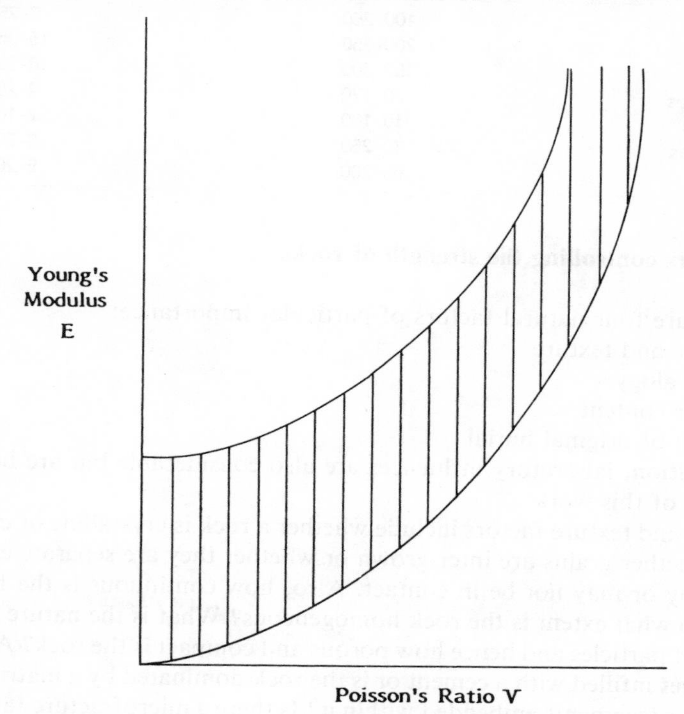

Fig. 2.18 Recorded ranges of Poisson's Ratio for values of Young's Modulus.

However, Poisson's ratio varies considerably in rocks with low values of Young's Modulus, i.e. non-elastic rocks (Fig. 2.18). For elastic solutions a Poisson's ratio of 0.25 is often taken unless there is overwhelming evidence to the contrary. Then, there are grounds for assuming an elastic behaviour and no value of Poisson's ratio would be required anyway.

2.11 Strength Properties of Rocks

In this context, strength is taken as uniaxial compressive strength and uniaxial tensile strength, which are the most commonly used strength values in civil engineering works. The variability of strength shown in Table 2.7 illustrates the problem of accurately determining only one of the parameters of the rock, the ranges of strength common in most types of rock being considerable.

Similar problems are faced in deciding on design values for other parameters e.g. Young's Modulus.

Table 2.7 Comparison of uniaxial compressive and uniaxial tensile strengths of rocks.

Rock Type	UCS (MN/m^2)	UTS (MN/m^2)
Granite	100-250	7-25
Dolerite	200-350	15-35
Basalt	150-300	10-30
Sandstones	20-170	4-25
Mudrocks	10-100	2-10
Limestones	30-250	5-25
Gneisses	50-200	5-20

1 Factors controlling the strength of rocks

There are four natural factors of particular importance:

1. Fabric and texture
2. Mineralogy
3. Water content
4. Depth of original burial

In addition, laboratory influences are also considerable but are beyond the scope of this work.

Fabric and texture factors include whether a rock is crystalline or clastic, that is whether grains are inter-grown or whether they are separate entities which may or may not be in contact. Also, how continuous is the fabric, that is, to what extent is the rock homogeneous? What is the nature of the grading of particles and hence how porous and compact is the rock? Are the pore spaces infilled with a cement or is the rock dominated by a matrix with the coarser fragments embedded within it? Is there a microfracture fabric in the rock which could be utilized to form a continuous failure plane? Many coarse-grained rocks which possess a fracture cleavage only reveal this characteristic under the microscope.

Quartz especially seems to impart strength to rocks (Fig. 2.19). Minerals with a pronounced cleavage, especially micas, tend to produce much weaker rocks. Secondary alteration in virtually all cases tends to produce minerals weaker than the original ones. Some minerals are chemically unstable and affect the strength of rock in service. Reactive silica would be a major problem in rocks used for concrete aggregates. The cement/matrix mineralogy in elastic rocks has a very significant bearing on strength, a matrix usually producing weak rocks, with increasing strengths through carbonate, iron oxide and silica cements.

Water content of rock generally shows the consistent influence of decreasing the strength of rock (Fig. 2.20). Normally, rocks are strength tested in the laboratory under dry or saturated conditions. Natural moisture

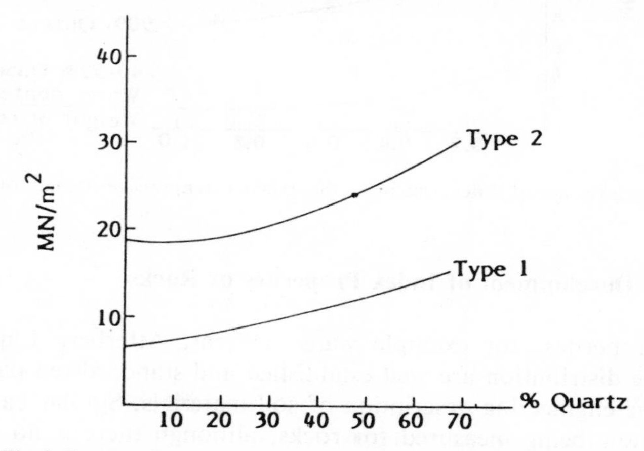

Fig. 2.19 The influence of quartz content on the uniaxial compressive strength of rocks.

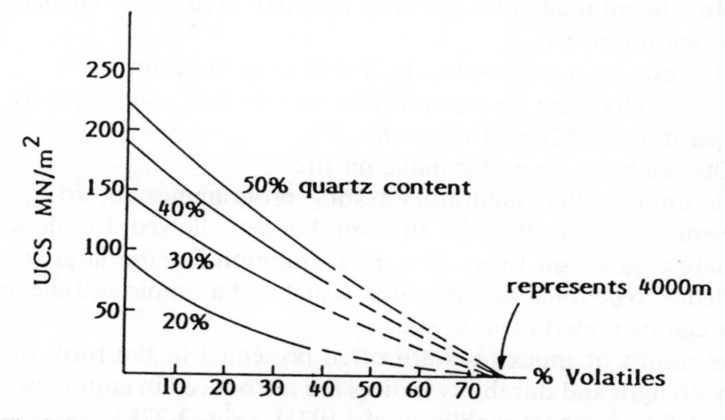

Fig. 2.20 The influence of depth of burial on the uniaxial compressive strength of rocks.

content tends not to assume the same level of importance as in soil mechanics.

The importance of depth of burial was discovered during tests on coal. Sandstones adjacent to the coals were tested, the water content and volatiles (gas) content of the coal decreasing with increasing depth (Fig. 2.21). The curves all appear to meet at a common point which confirms the relationship.

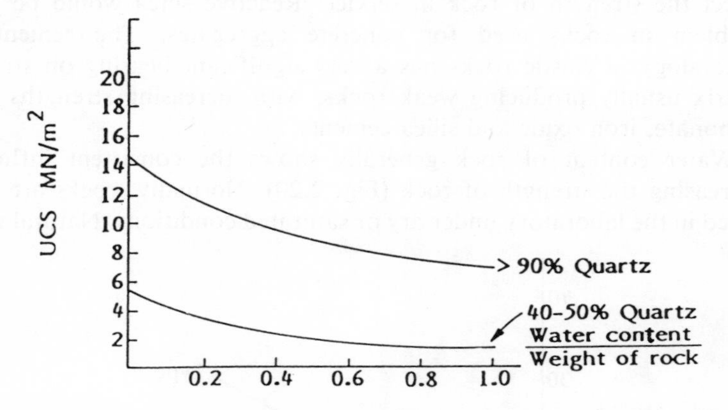

Fig. 2.21 The influence of water content on the uniaxial compressive strength of rocks.

2.12 The Development of Index Properties of Rocks

Basic properties, for example water content, Atterberg Limits, and particle size distribution are well established and standardized parameters used for the engineering description of soil materials. Similar basic properties are now being measured for rocks, although there is no universal agreement of what constitutes index properties for rocks. As a guideline, the tests adopted should meet the following criteria:

1. Must be an index of a material property used by an engineer to solve design problems.
2. The tests must be simple, cheap and easy to perform.
3. Test results must be reproducible on standard equipment by different operators in different locations.
4. The test must be performable on site.

It is unlikely that laboratory testing programmes for rocks will yield sufficient results within the time or budget allowed for designs to be adequately based on them. It is now common for the large gaps in this knowledge to be filled by index tests, usually of a simple and cheap method, which can be undertaken on site.

The results of index tests are often presented in the form of a log, in which strength and durability indices are introduced in conjunction with an index of fracturing, (Franklin *et al.*, 1971), (Fig. 2.22).

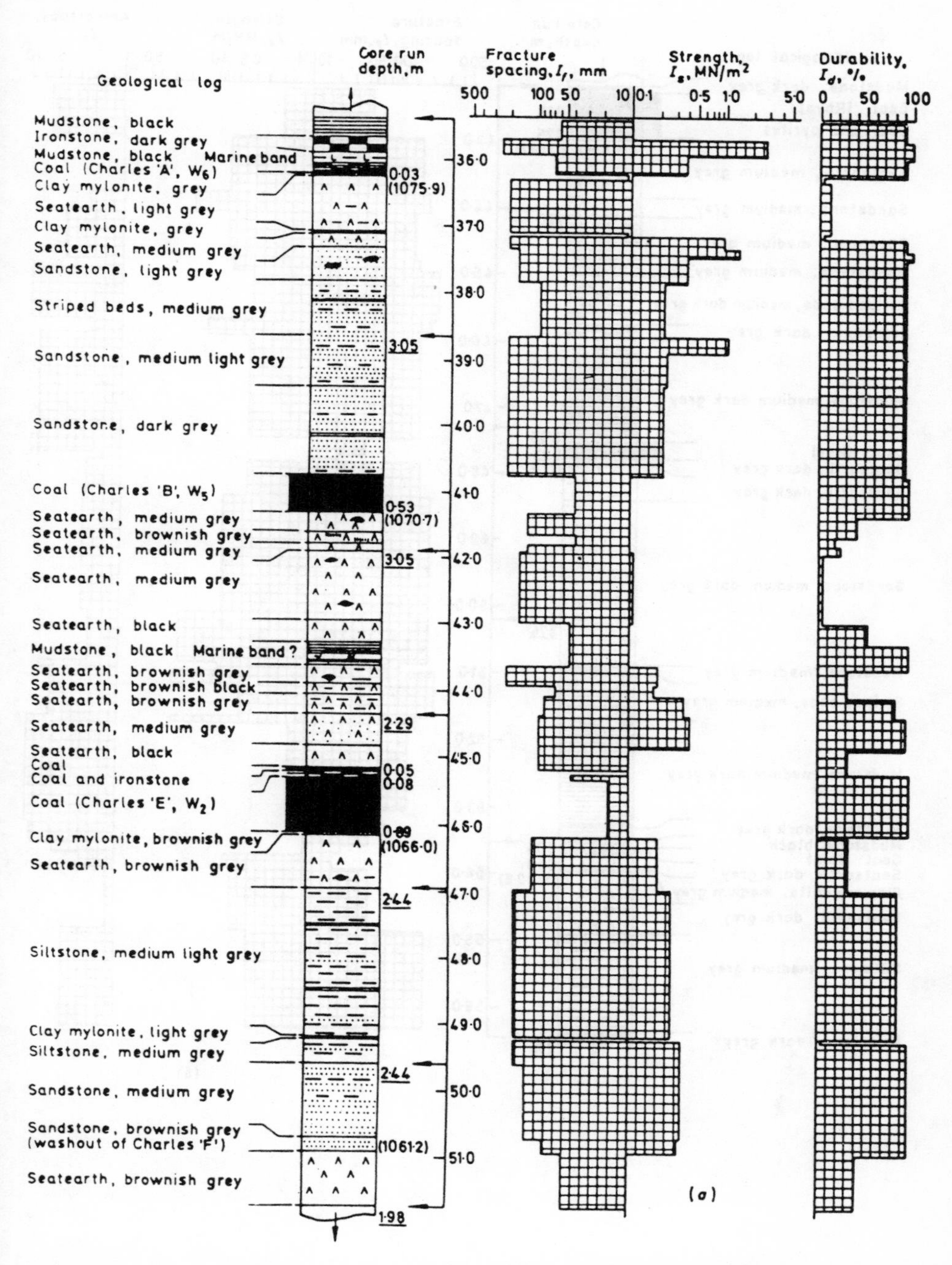

Fig. 2.22 Rock core logs showing variations in some basic mechanical properties. (Franklin, J.A., Broch, E. and Walton, G., 1971; Copyright © 1971 by *Institution of Mining and Metallurgy*)

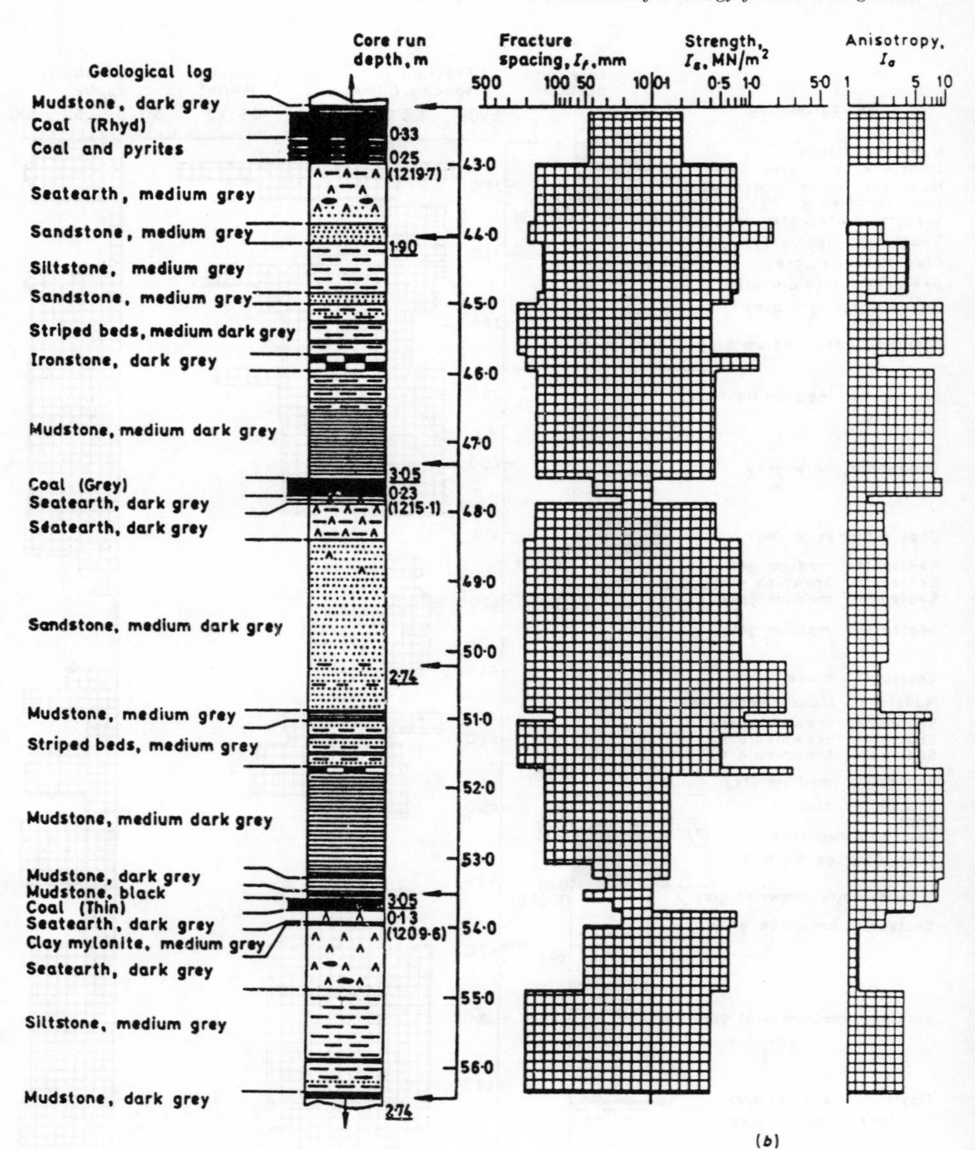
Geological log
Core run depth, m
Fracture spacing, I_f, mm
Strength, I_s, MN/m^2
Anisotropy, I_a
500 100 50 10 0·1 0·5 1·0 5·0 1 5 10
Mudstone, dark grey
Coal (Rhyd)
Coal and pyrites
Seatearth, medium grey
Sandstone, medium grey
Siltstone, medium grey
Sandstone, medium grey
Striped beds, medium dark grey
Ironstone, dark grey
Mudstone, medium dark grey
Coal (Grey)
Seatearth, dark grey
Seatearth, dark grey
Sandstone, medium dark grey
Mudstone, medium grey
Striped beds, medium grey
Mudstone, medium dark grey
Mudstone, dark grey
Mudstone, black
Coal (Thin)
Seatearth, dark grey
Clay mylonite, medium grey
Seatearth, dark grey
Siltstone, medium grey
Mudstone, dark grey
0·33
0·25
(1219·7)
1·90
3·05
0·23
(1215·1)
2·74
3·05
0·13
(1209·6)
2·74
43·0
44·0
45·0
46·0
47·0
48·0
49·0
50·0
51·0
52·0
53·0
54·0
55·0
56·0

(b)

CHAPTER 3: THE GEOMETRY, DESCRIPTION AND PROPERTIES OF ROCK MASSES

Rock structure is a term used to describe the overall relationship of rock masses, for example folding, jointing, cleavage, and unconformities. Folding of rock changes the orientation of rocks by lateral compression, the changes perhaps being most apparent in bedded rocks. The compressive forces causing folding, and indeed fracture structures as well, originate most usually at destructive plate margins. The tilting of rocks from their original horizontal position, and the fracturing of rocks, produces measurable directional properties in the rocks called strike and dip (Fig. 3.1). The strike is the direction at any point on a structural surface of a horizontal line drawn on the surface. The term is also used in the sense of the general trend of rocks. It is at right angles to the true dip. True dip is the vertical angle measured from the horizontal plane in the direction of greatest slope. The apparent dip is the dip in any direction from zero in the direction of

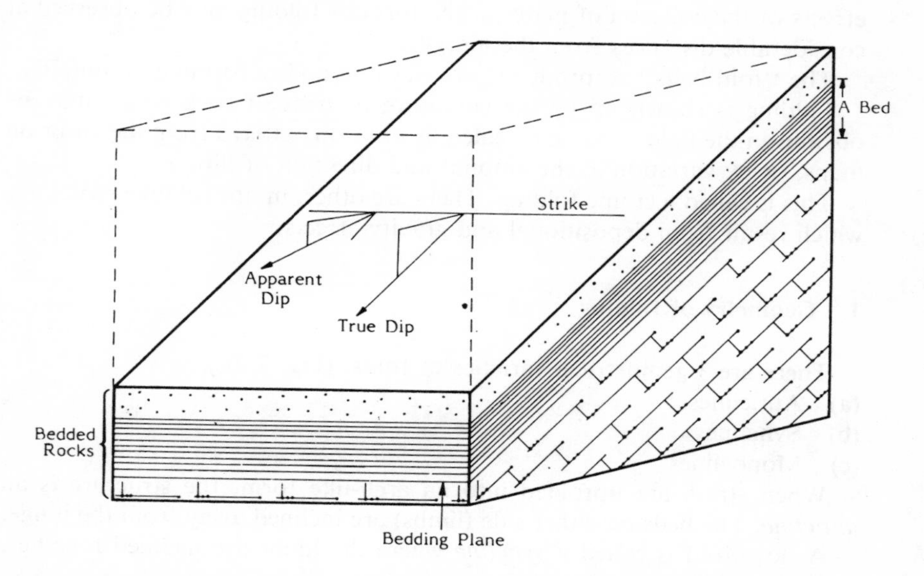

Fig. 3.1 Illustration of the form and basic directional properties of bedding planes.

strike to a maximum in the direction of true dip. The measurement of these properties by geologists is in terms of strike, measured as a whole compass bearing from north (for example, a bed striking NE-SW would have a strike of 045 ° or 225 ° — they are exactly the same), a dip angle, which is self explanatory, and a general dip direction, usually as one of the eight basic cardinal points of the compass. So, in Fig. 3.1, if north is in a horizontal plane passing directly into the page the bed illustrated would have the following approximate orientation.

Strike 290 °, dip 35 °, dip direction, S.W.

In engineering geology, it is more usual to measure true dip, and dip direction, the latter as an accurate whole compass bearing. So, the plane in Fig. 3.1 would become:

dip 35 °, dip direction 200 °

Note that the dip direction is 90 ° from the strike.

3.1 Folds

The earth's crust may undergo deformations of various kinds. Deformation belts are relatively narrow, long zones of originally flattish sediments that became crumpled, disturbed, often metamorphosed and intruded. They occur most commonly at destructive plate margins, although the effects of the collision of plates in the form of folding may be observed at considerable distances from the margins.

The "fold belts" so produced are elevated bodies forming mountains.

Folding is shown up by the curvature of beds of rock which may be obvious in the field, may be revealed by outcrop patterns from the air or on maps, or by variation in the amount and direction of dip.

This is called tectonic folding. There are other, minor foldlike structures which result from depositional and gravity effects.

1 Geometry of folding

There are 3 geometrical varieties of folds: (Fig. 3.2)

(a) Anticlines
(b) Synclines
(c) Monoclines

When strata are upfolded into an arch-like form, the structure is an *anticline*. The beds on either side (limbs) are inclined away from the hinge.

A downfold is called a *syncline* where the limbs are inclined together, toward the hinge.

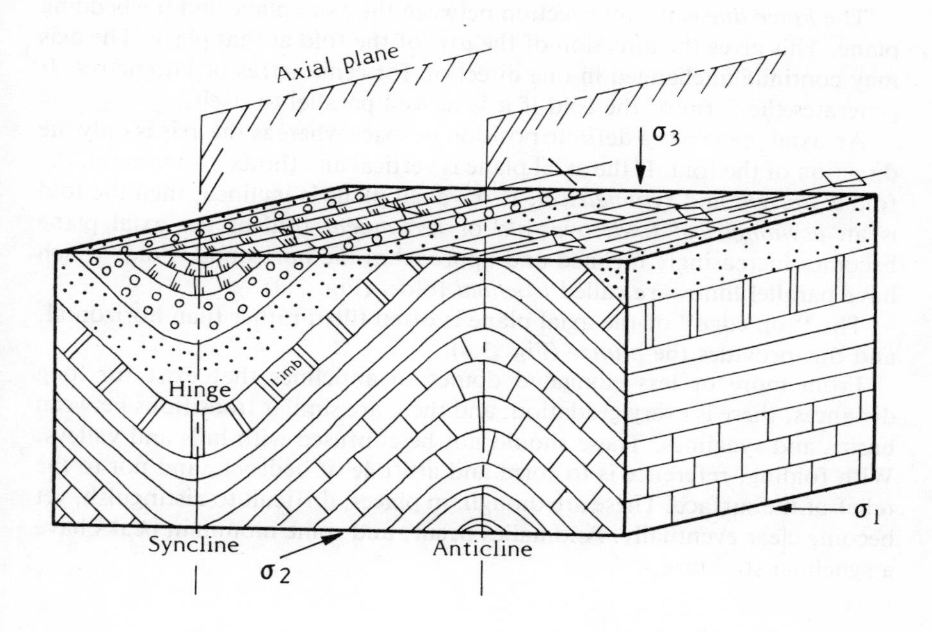

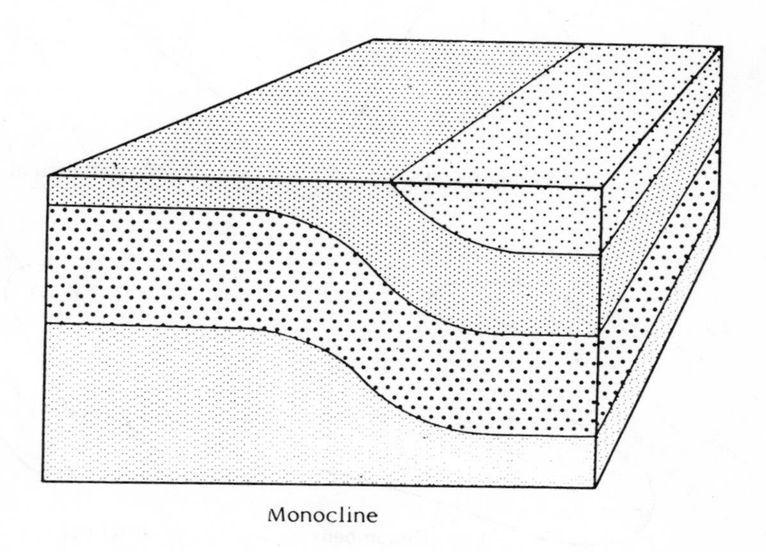

Fig. 3.2 The form and properties of simple folds.

The *axial plane* of the fold bisects the angle between the two limbs. It is defined in terms of strike and dip.

The *hinge line* is the intersection between the axial plane and the bedding plane. This gives the direction of the *axis* of the fold at that place. The axis may continue unchanged in one direction for centimetres or kilometres. It generates the form of the fold if it is moved parallel to itself.

An axial plane has a definite position in space whereas the axis is only the direction of the fold. If the axial plane is vertical and the axis horizontal, the fold is *upright* and *symmetrical*. If the axial plane is inclined, then the fold is an *asymmetrical fold*, *overfold* or *recumbent* fold as the axial plane becomes increasingly inclined and approaches the horizontal. Folds which have parallel limbs are called *isoclinal* folds. (Fig. 3.3).

The "top edge" of the axial plane is often tilted rather than horizontal, and this provides the *plunge* (Fig. 3.4).

From more or less elongated domes to anticlines that exist for long distances, there is every gradation; and there are similar transitions between basins and synclines. These should not be confused with hills and valleys. With folding, reference is to form and attitude of bedrock, and not to the relief of the surface. These are though, in places, difficult to distinguish, yet become clear eventually. Anomalies occur, and some mountain peaks have a synclinal structure.

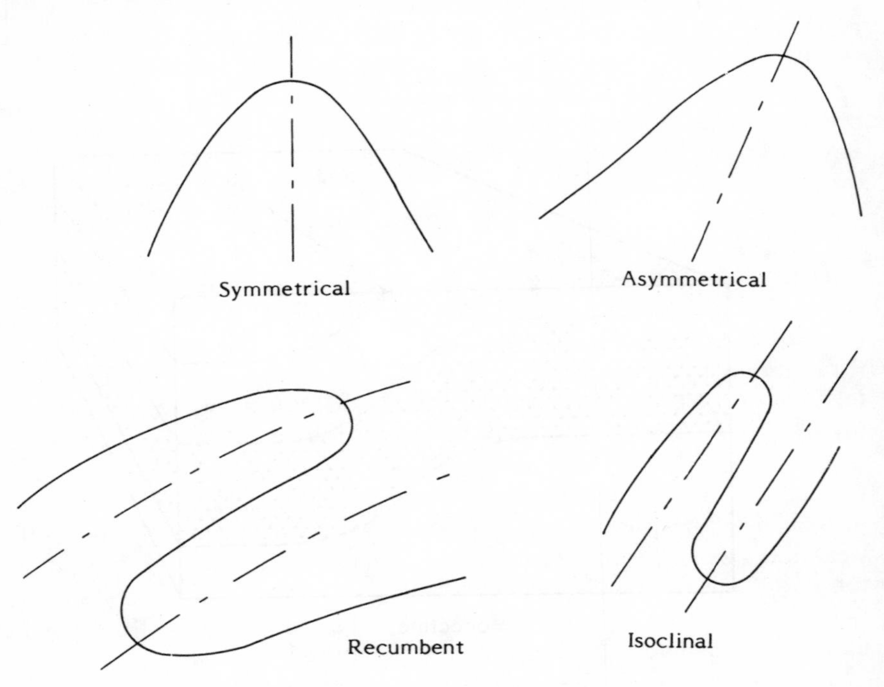

Fig. 3.3 Variations in the geometry of folds.

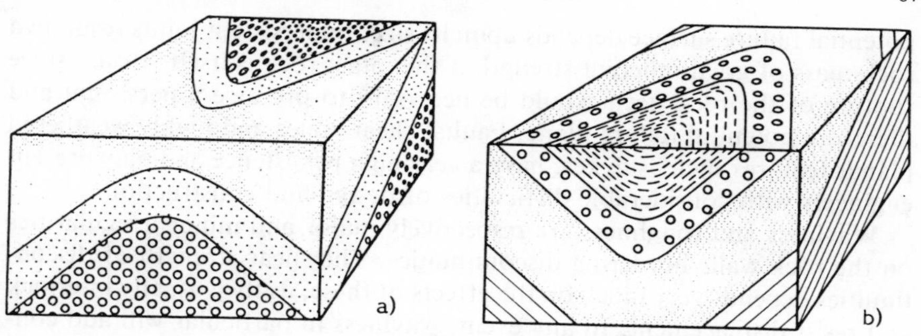

Fig. 3.4 The effects of plunge on outcrop patterns; a. plunging anticline, b. plunging syncline.

3.2 Structural Discontinuities

Discontinuity is the term used in engineering geology for a break, fracture, or sudden change in a rock mass. It usually refers to faults, joints, cleavages, bedding planes, and other fractures, but may be extended to cover veins and even dykes. Fracture discontinuities will be dealt with here.

Structural discontinuities are universally present in rock masses, irrespective of their origins. Such features have appreciably lower strengths than the intact rock, and much of the strength and stability of rock masses depends on the strength of discontinuities rather than the intact rock. Discontinuities impart anisotropy onto rock masses, because most fractures have a tectonic origin, and hence a certain uniformity of orientation. So, discontinuities often occur in groups, in joints generally referred to as sets, and any of these fractures may be observed with respect to

(a) attitude (orientation)
(b) geometry
(c) spatial distribution

Depending upon the origin of the discontinuities, their characteristics can vary greatly with respect to

(a) spacing (number per metre)
(b) infilling materials (gouge)
(c) physical characteristics of the planes (waviness or roughness)
(d) degree of development (how far they can be traced — persistence)

Spacing or intensity will partly indicate the extent to which the intact rock and the discontinuities will separately affect the mechanical properties of the mass. Rock masses with a high fracture frequency will be inherently weaker than those with a low fracture frequency. However, the nature of the development of the discontinuities is also important in that minor, randomly oriented features will be generally of less significance than uniformly orientated structures.

Persistence is the most difficult feature of discontinuities to determine, but it is essential that it is measured because the strength reduction on a potential failure surface depends upon its size. Impersistent joints result in a rock mass of great inherent strength if the intact rock is itself strong, since rupture of the rock itself would be necessary to produce a persistent and therefore potential failure plane. Faults, shear zones and highly weathered planar features such as dykes, have a very high persistence and may present engineers with considerable difficulties of design and construction.

Waviness and roughness are respectively major and minor irregularities on the rock walls bordering discontinuities. The shear strength of discontinuities depends very much on the effects of these, as they usually interlock and resist displacement. In any event, waviness in particular will add considerably to the angle of shearing resistance to be expected. In general, discontinuities which have a tensional origin, e.g. tension joints, bedding planes, cooling joints, will have greater roughness than ones having their origin in shear.

In general terms, gouge includes any material that occurs between two structural planes which is different from the host rock. In faults this may be broken or ground-up rock, clay, a mineral vein, etc. In certain cases, the gouge may be washed in, for example in limestones, where solution has enlarged a discontinuity and clay, or some surface material has been washed down by rainfall. The effect of gouge on strength properties is very great, and the strength varies between the strength of the gouge itself, where the thickness of gouge is sufficient to keep apart the rock walls during shear, through all ranges of modifications to the angle of shearing resistance, to a strength depending purely on the rock walls where no gouge exists. A classification of filled discontinuities was given by Barton (1974) and is shown in Fig. 3.5.

1 Faults

Faults are probably the major type of structural discontinuities, although they are also the least frequent. Faults are major breaks in the earth's crust producing displacements which range from vertical to horizontal depending on the fault type. The major types of faults are (Fig. 3.6):

1. Normal dip-slip faults
2. Reverse dip-slip faults
3. Low-angle faults (thrust and lag faults)
4. Strike-slip faults (wrench faults)

As was mentioned in the section on structural discontinuities, faults are of major significance in civil engineering. They are long features of sometimes major proportions. They are often infilled with weak, deeply weathered gouge, and are often major sources of groundwater. Not all faults are of such a large magnitude, but their scale is virtually always larger than other fractures, and they are generally expected to have smooth rock walls, the waviness and roughness having been sheared off during the formation of the fault.

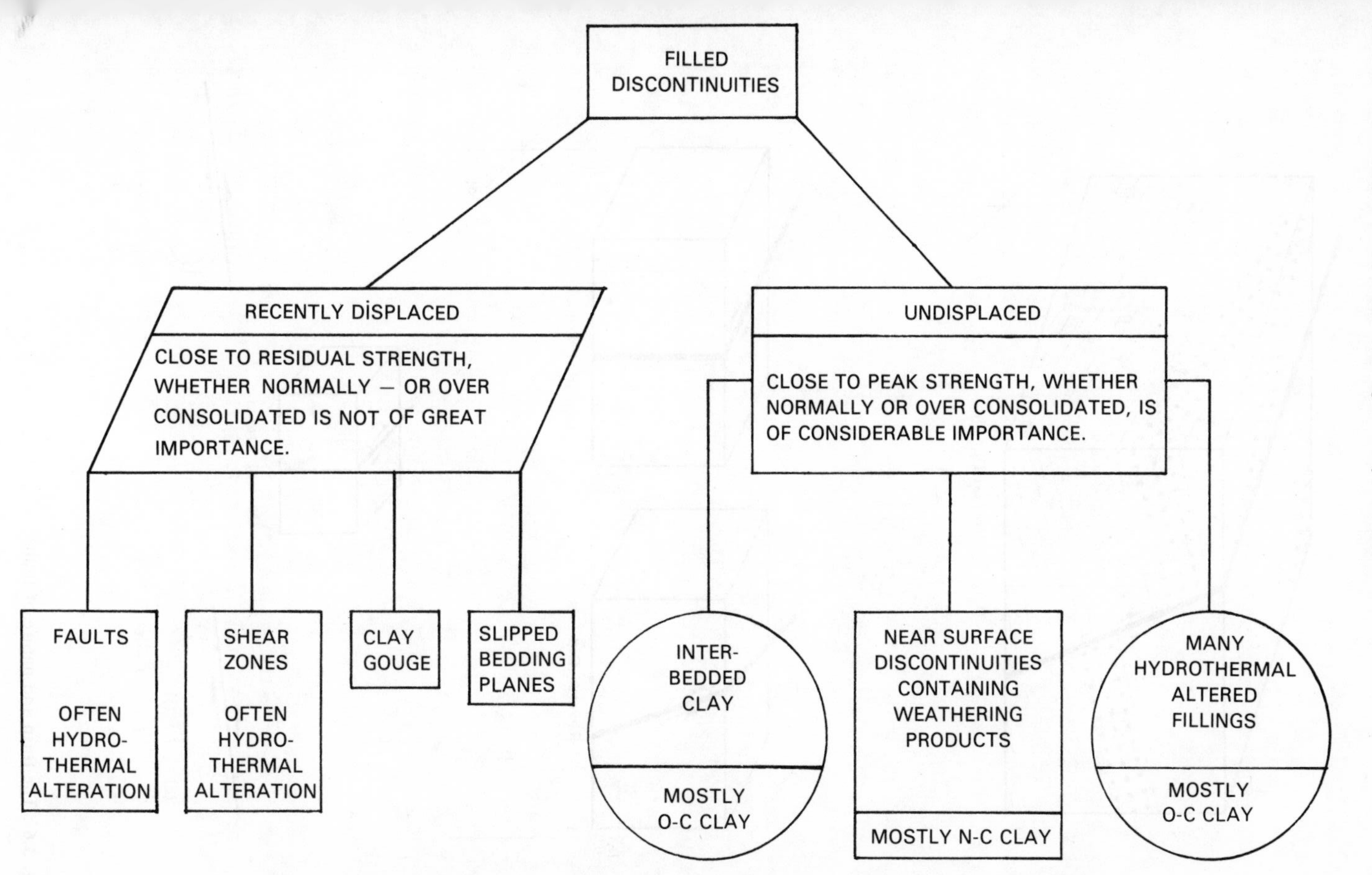

Fig. 3.5 Classification of filled discontinuities. (Barton, N., 1974; Copyright © 1974 by *Norwegian Geotechnical Institute*)

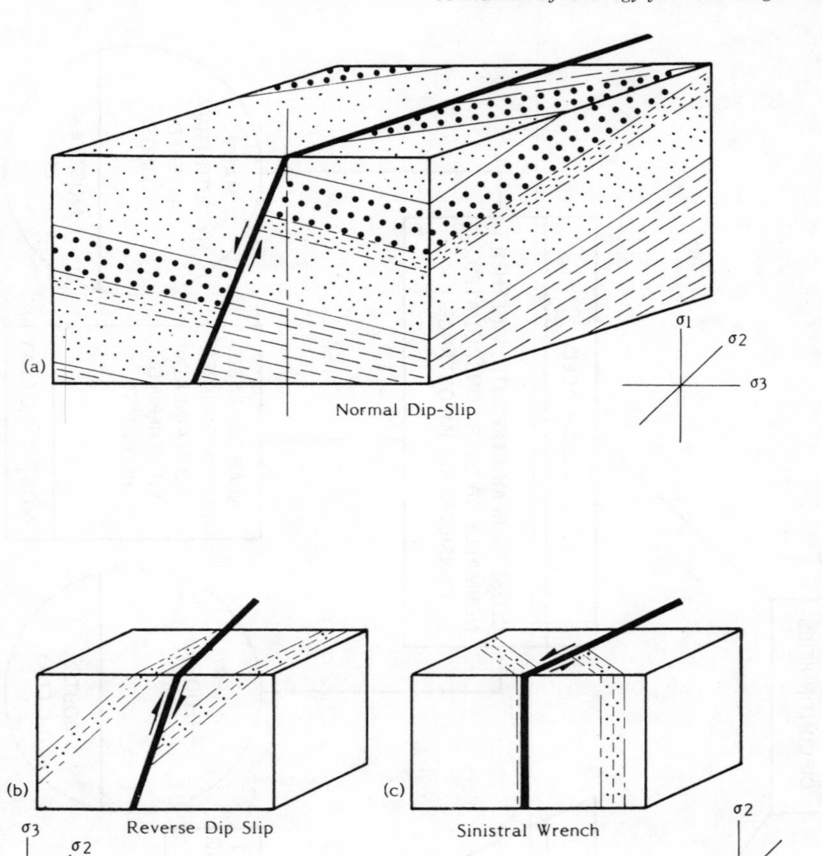

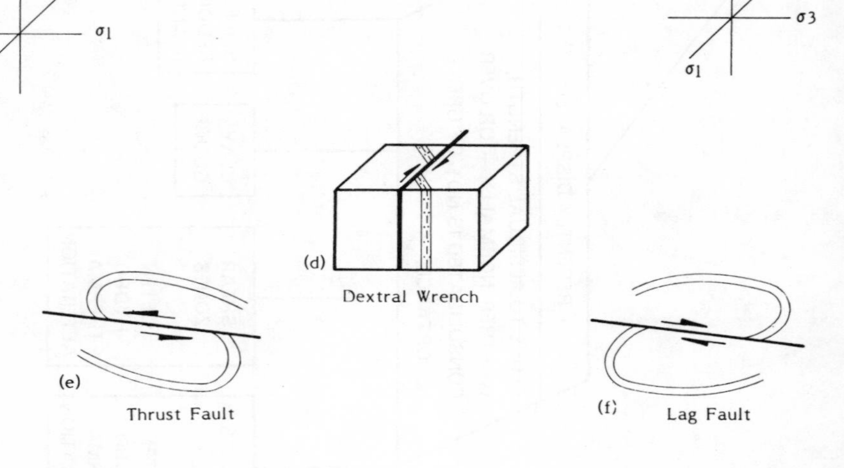

Fig. 3.6 The form and nature of faults.

2 Rock cleavage

In geology, cleavage applies to a structure which permits a regular splitting of rocks. Unfortunately, classifications of cleavage are extremely confused in that both genetic and descriptive criteria may be employed in the same classification. Furthermore, there is often disagreement as to precisely how some cleavages do develop. Powell (1979) has introduced a comprehensive classification based on morphological characteristics, although a much simplified version is generally adequate when describing cleaved rocks in engineering terms. The basis of classification here is a simplified genetic one, and will consist simply of two broad types (Fig. 3.7):

1. Cleavage due to rotation of mainly platey minerals and therefore essentially continuous and smooth.
2. Cleavage due to rotation of massive minerals or crystals, and therefore essentially non-continuous and rough.

The nearest equivalents to these types in general terminology would be slaty cleavage, and fracture cleavage. In rocks composed of platey minerals, whether they are sedimentary or metamorphic, reorientation of the minerals normal to the maximum principal stress, with or without recrystallization, usually results in a continuous, closely spaced fracture fabric. In slates, this is likely to be extensive throughout the whole rock mass. In mudrocks, it will largely depend on the thickness of the bed, but this type of cleavage is unlikely to be extended into, for example, adjacent sandstones. The cleavage may develop as a result of stressing during regional metamorphism, or parallel to the axial plane of folds during phases of folding.

Cleavages which develop in coarser rocks composed of massive grains or crystals, tend to be less "perfect". In crystalline rocks, most cleavage probably results from fracture of crystal boundaries in their attempt to rotate during stressing. In clastic rocks, fracturing of the cement or matrix material

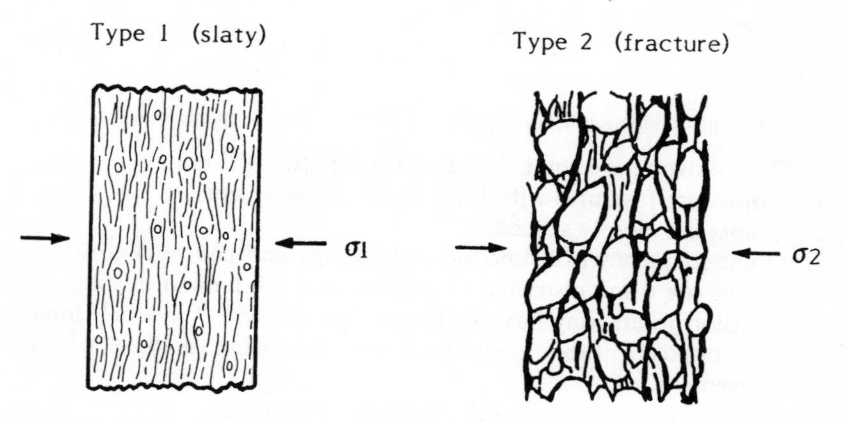

Fig. 3.7 The form of the two basic types of rock cleavage.

during the rotation of grains is probably responsible. Therefore, to a large extent, it is the orientation of the small fractures rather than directly of the grains or crystals, which accounts for the formation of the second type of cleavage. In the field, this cleavage is often not visible, but can be clearly seen in thin section. This may lead to over confidence in the strength of the rock which is likely to have a major directional weakness along the fracture cleavage.

3 Joints

Joints may occur either through tensional or compressive forces. They are commonly associated with other features of tectonic deformation (Fig. 3.8), but may also result from cooling of igneous rocks, or shrinkage of sediment during diagenesis. Joints also frequently occur as a result of the release of stored stresses in rock which accumulate mainly during phases of tectonic compression, but also during cooling and shrinkage of igneous rocks, and compression of sedimentary rocks beneath overburden.

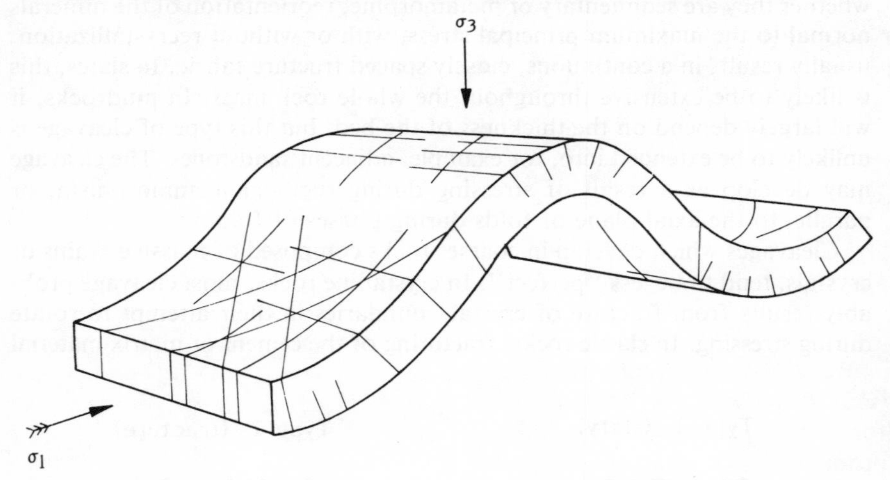

Fig. 3.8 The distribution of joint planes which result from simple folding.

The following are some of the characteristics of joints.

1. Joints are fractures with little or no displacement.
2. Joints are closely spaced.
3. Joint frequency is a function of lithology and bed thickness.
4. Joints are often restricted to one bed and are limited in size.
5. Tension joints may have surface roughness with "mirror" images on the opposite face, their interlock indicating a lack of relative movement.
6. Joints are often in parallel sets in any one area.
7. Joints are usually oriented symmetrically and normal to bedding.

4 Unconformities

Several types of unconformity exist, but there are two major types. Angular unconformities occur where there is a primary discordant structural relationship between one rock body and the one laid down on top of it, that is, the dip and strike directions of the two rock series are different (Fig. 3.9). These are usually planar since erosion of the upper surface of the older rocks has normally been by the sea. The orientation of the upper series of rocks is the same as that of the plane of unconformity. Very often, the bed immediately above the unconformity is a conglomerate.

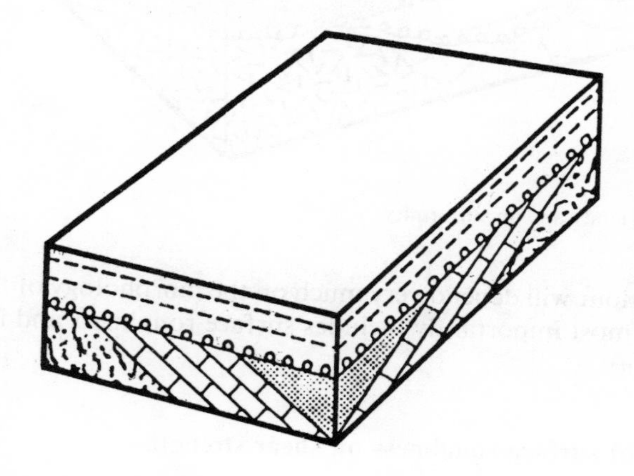

Fig. 3.9 Angular unconformity.

Buried landscape unconformities may or may not have significant structural differences between the rock formations. In this case, an older formation has been weathered and eroded by normal terrestrial agencies, producing hills and valleys, etc. A later set of rocks is then laid down completely covering the irregular surface. This most frequently occurs in desert conditions where pre-existing rocks become wind eroded or gullied, and wind blown sand, or other deposits cover the features over. (Fig. 3.10).

Unconformities also include contacts between igneous intrusions and later rocks.

3.3 Shear Strength of Discontinuities

Discontinuities occur in all rock masses and represent pre-existing lines of weakness which to a greater or lesser extent will determine the overall behaviour of the mass. From a shear strength point of view, discontinuities assume particular significance in excavations of all kinds. Details of the

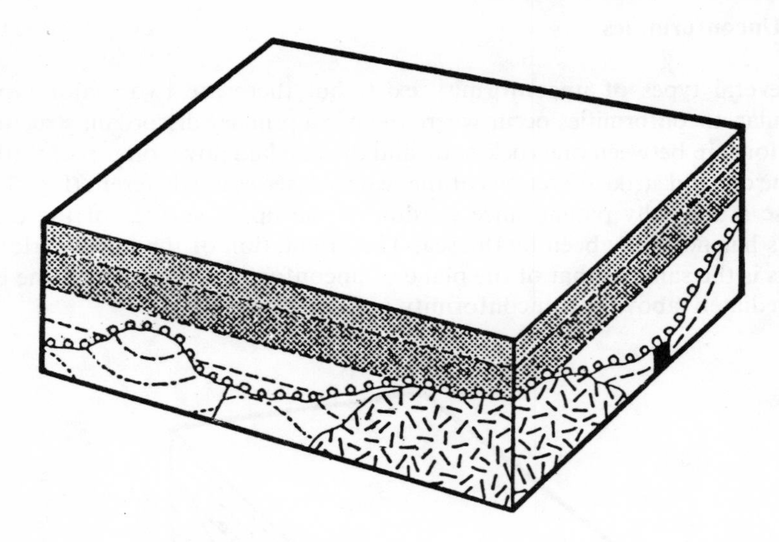

Fig. 3.10 Buried landscape unconformity.

shearing behaviour will depend very much on the morphology of the discontinuity, which most importantly includes surface roughness and infilling of the discontinuity.

1 Influence of surface roughness on shear strength

If the specimen shown in Fig. 3.11a is subjected to shear and normal loads, shear can take place by
(1) over-riding of the projections
(2) shearing through of the projections

In the case of over-riding, displacement is no longer parallel to the shear stress (Hoek and Bray, 1981), (Fig. 3.11b). Assuming that the surface has zero cohesion, the strength of the rough joint would be (Patton, 1966):

$$S = \sigma_n \tan(\phi + i)$$

where S is the shear strength along the joint
σ_n is the normal stress
ϕ is the basic angle of shearing resistance of the material
i is the angle of the roughness measured from the horizontal.

If dilation is inhibited and shearing through of the projections occurs, (Fig. 3.11c), the resulting fracturing of the intact rock will produce an apparent cohesion, and its strength may be described as

$$S = c + \sigma_n \tan \phi .$$

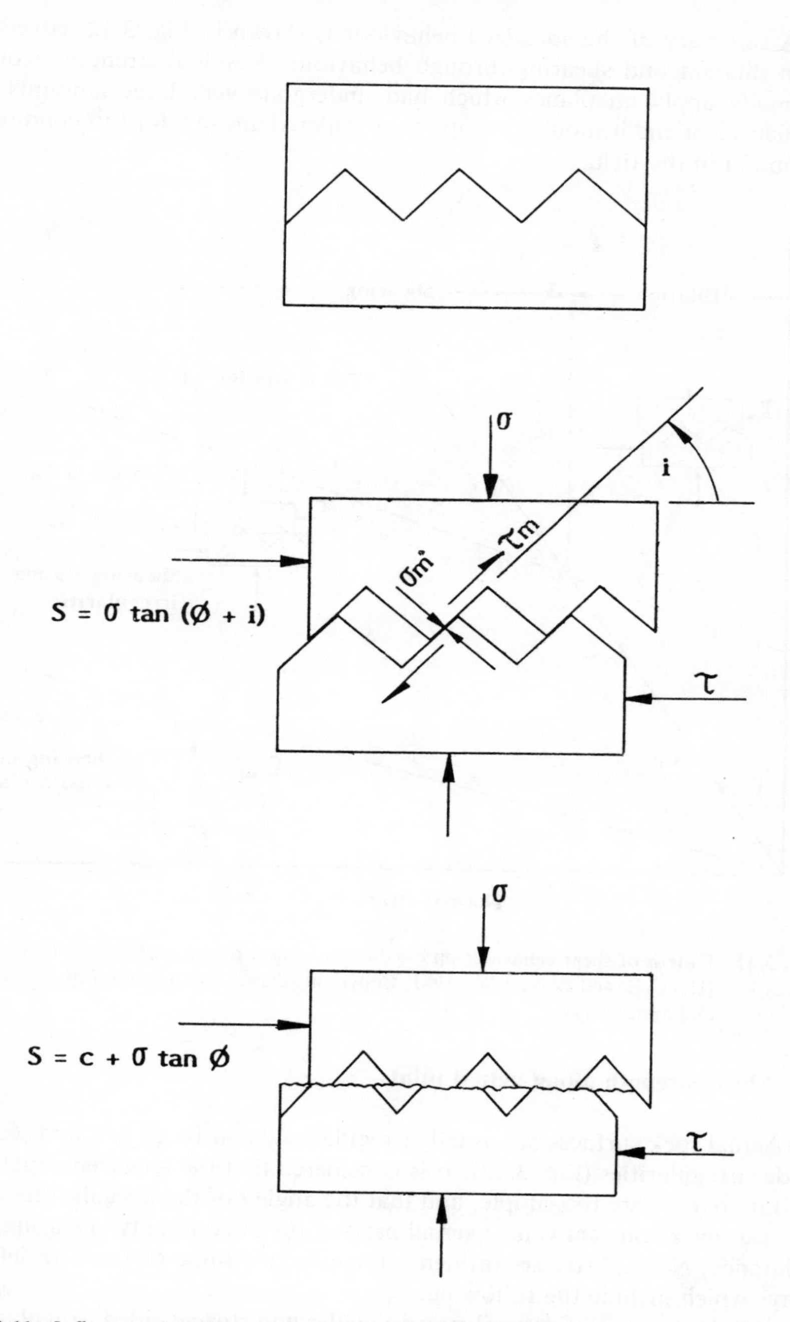

Fig. 3.11 Influence of surface roughness on the shearing behaviour of discontinuities. (Hoek, E. and Bray, J.W., 1981; Copyright © 1981 by *Institution of Mining and Metallurgy*)

A summary of this idealized behaviour is shown in Fig. 3.12, covering both dilatant and shearing-through behaviour. Residual strengths would normally apply on planes which had undergone very large amounts of displacement and it should perhaps be considered unusual for this condition to obtain in the field.

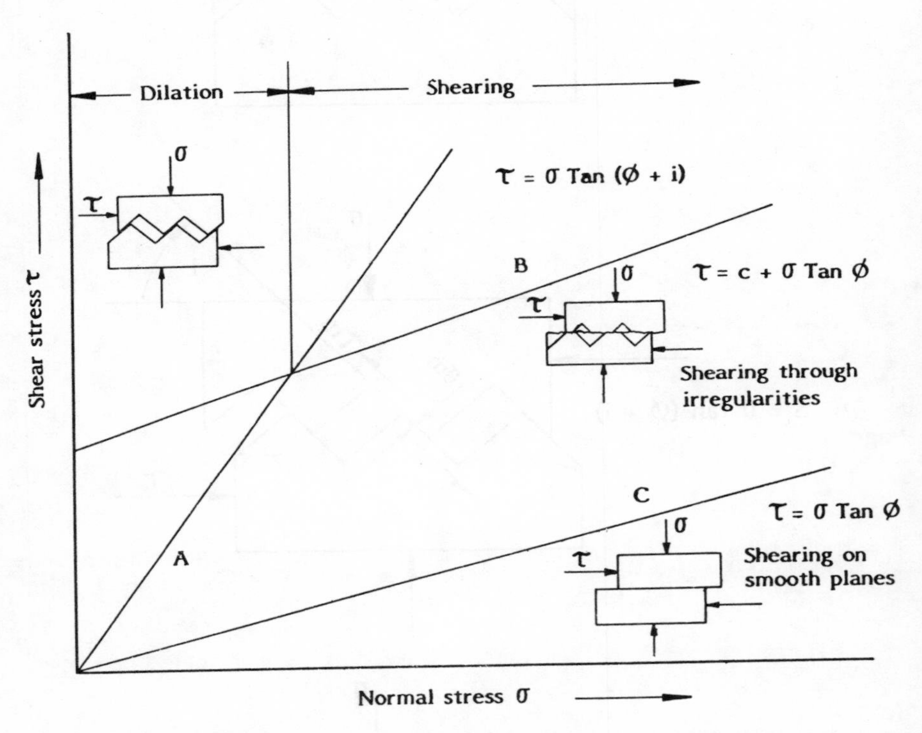

Fig. 3.12 Pattern of shear behaviour on discontinuities as a function of morphology. (Hoek, E. and Bray, J.W.. 1981; Copyright © 1981 by *Institution of Mining and Metallurgy*)

2 Shear strength along actual joints

Actual rock surfaces are usually irregular and consist of first and second order irregularities (Fig. 3.13). It is considered that the idealized situations so far covered are too simple, and that the angle *i* of the irregularities does not assume a constant value over all parts of the discontinuity or throughout dilatancy. Normal stresses influence the effective value of *i* in a variety of ways which include the following.

(a) At low values of normal stress, smaller and steeper sided irregularities control movements.

(b) At higher values of normal stress, the smaller projections are broken

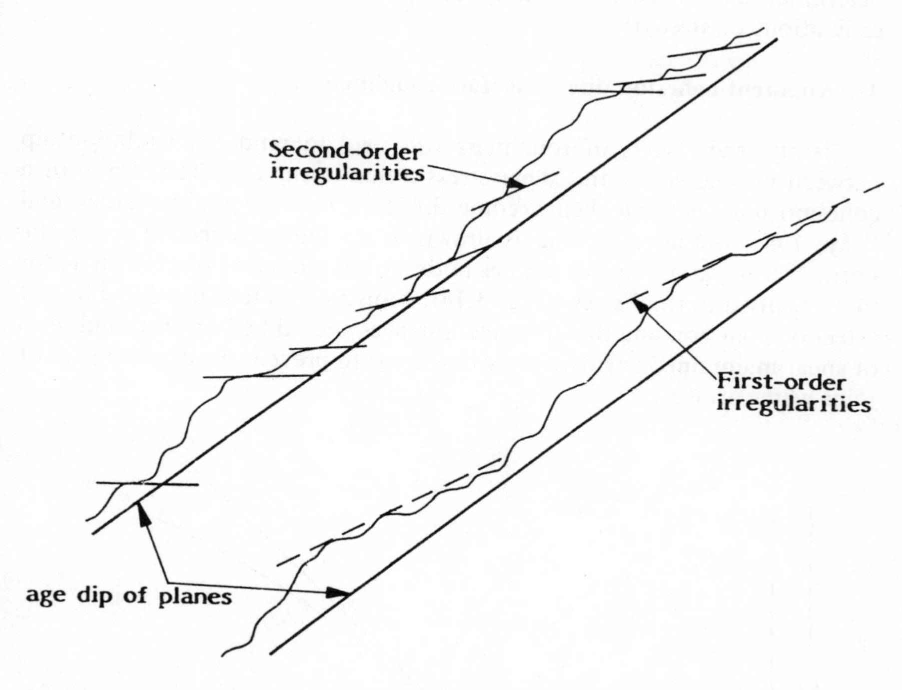

Fig. 3.13 First and second order irregularities on discontinuity surfaces. (Hoek, E. and Bray, J.W., 1981; Copyright © 1981 by *Institution of Mining and Metallurgy*)

off, and the primary "waviness", which has a lower angle i becomes the controlling influence.

(c) At high levels of normal stress, all roughness is sheared off.

3 Determination of *i*

Although it is possible to measure i directly, the variation which is likely to occur on any discontinuity makes it difficult to be sure that the value chosen is representative. The following empirical relationship has been offered (Barton, 1974) for the determination of representative values, clearly emphasizing the stress dependence of the parameter.

$$i = 20 \log_{10} (\sigma_c / \sigma),$$

where σ is normal stress (choose a number of values) and σ_c is the unconfined compressive strength of the wall material of the discontinuity. This may be measured with a Schmidt Hammer so that the effects of weathering can be taken into account. Point Load Strength determinations on core or locally obtained samples would lead to over estimations of strength.

4 Apparent cohesion due to surface roughness

Clearly, the effects of roughness will tend to render the relationship between normal stress and shear stress a non-linear one. The choice of a cohesion intercept can then become difficult. One suggestion (Hoek and Bray, 1981) is that a tangent is drawn to the failure curve at a specific normal stress, and that this defines both the cohesion and friction angle for that specified normal stress (Fig. 3.14). Moisture content has a significant effect on cohesion and therefore test surfaces should be kept wet, and rates of shear maintained at sufficiently low levels to prevent the development of joint water pressures.

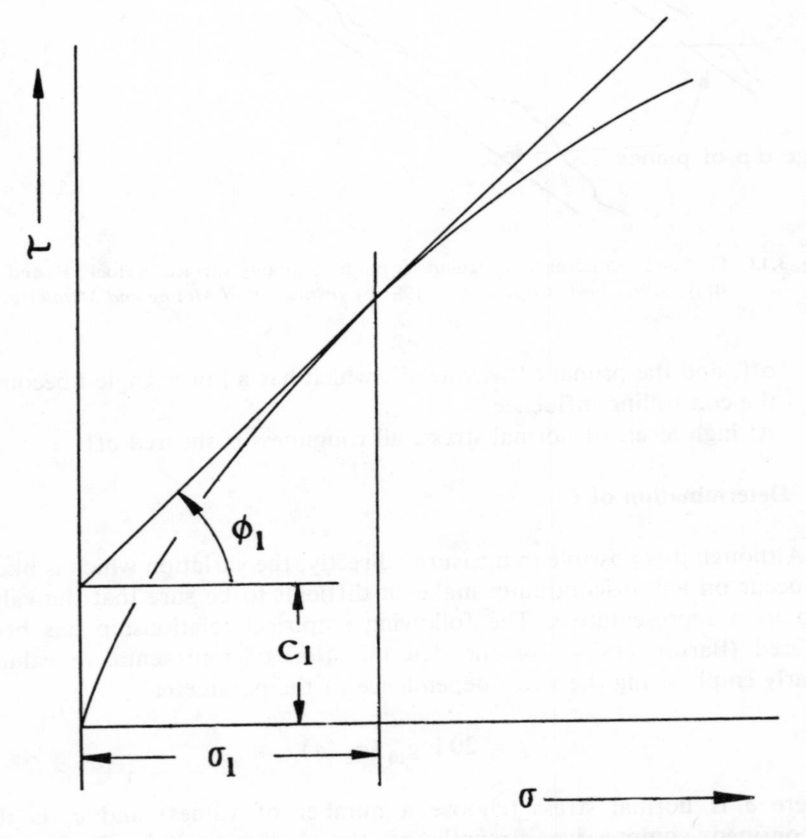

Fig. 3.14 The determination of cohesion on non-linear Mohr-Coulomb envelopes in rocks. (Hoek, E. and Bray, J.W., 1981; Copyright © 1981 by *Institution of Mining and Metallurgy*)

5 Peak and residual strength

Shear strength test results are shown for a typical discontinuity in a hard rock (Fig. 3.15). At a constant normal stress level, the shear stress required to displace the discontinuity increases very rapidly for the first 5 to 10 mm of displacement. This results from the interlocking of irregularities.

Eventually, the shear stress exceeds the strength of the joint, and further displacement occurs without change in shear stress. The high limiting value is called the PEAK shear strength. After a displacement of 50 mm or some-

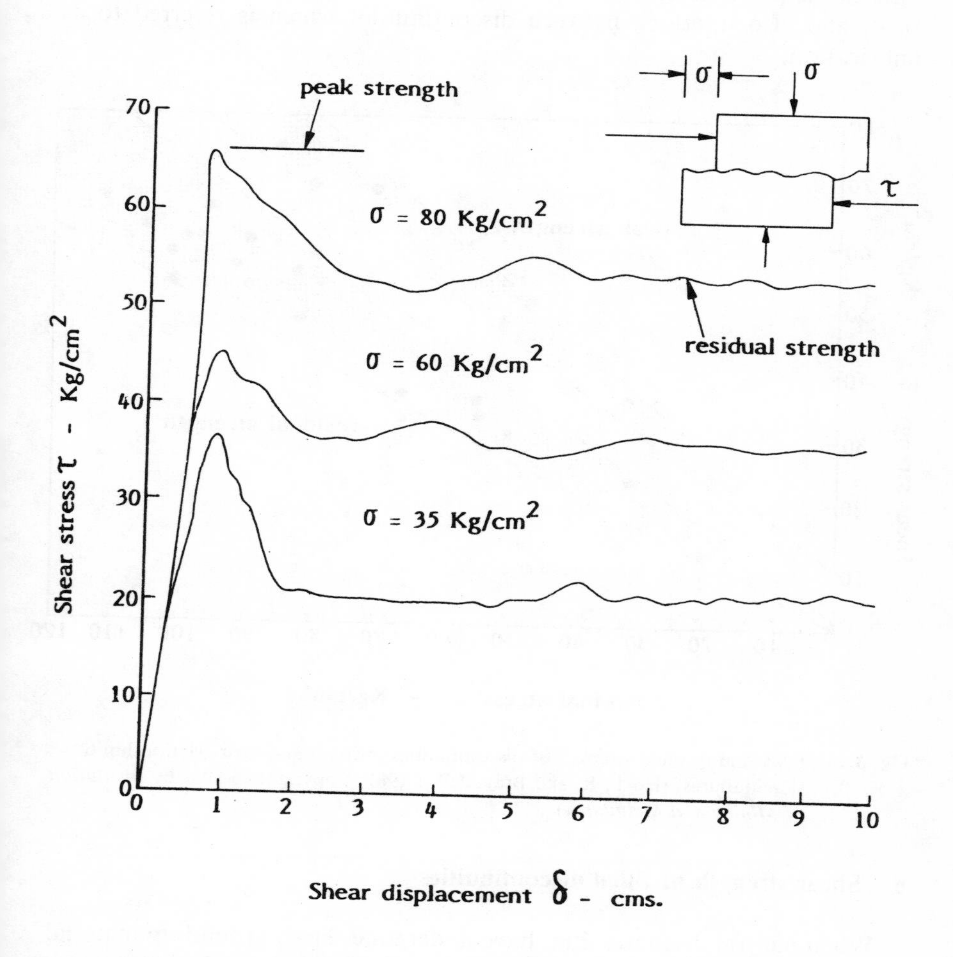

Fig. 3.15 Patterns of displacement during shear strength tests on discontinuities in rocks. (Hoek, E. and Bray, J.W., 1981; Copyright © 1981 by *Institution of Mining and Metallurgy*)

times more, gouge material has been formed from the ground down irregularities. This covers the surface and becomes polished and slickensided, resulting in the residual or ultimate strength of the discontinuity. However, it should be noted that irregularities in the displacement curve still exist and that very great displacements are required to produce a truly smooth surface. The ranges of peak and residual strength are shown for shear strength tests on clean joints in a single rock type (Fig. 3.16).

Investigations into precisely what shear strength parameters should be employed in, for example slope design, strongly indicate that failures occur at values close to the residual strength. Deviation from this minimum value reflects the presence of some surviving roughness on natural planes in the field, and also interlock between discontinuities which is referred to as imbrication.

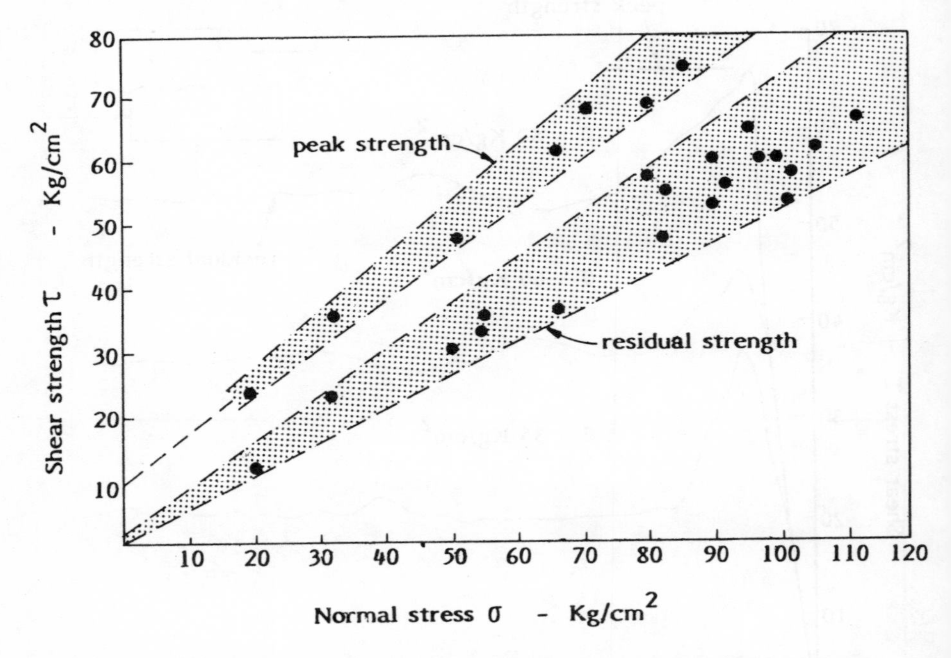

Fig. 3.16 Peak and residual strength of discontinuities in rocks and their relationship to slope failures. (Hoek, E. and Bray, J.W., 1981; Copyright © 1981 by *Institution of Mining and Metallurgy*)

6 Shear strength of filled discontinuities

When natural discontinuities have undergone shear, ground-up material or recrystallized material is frequently found on the plane. In other cases, material may be washed in and emplaced along the discontinuity. These infilling materials are called gouge, and they may have a considerable

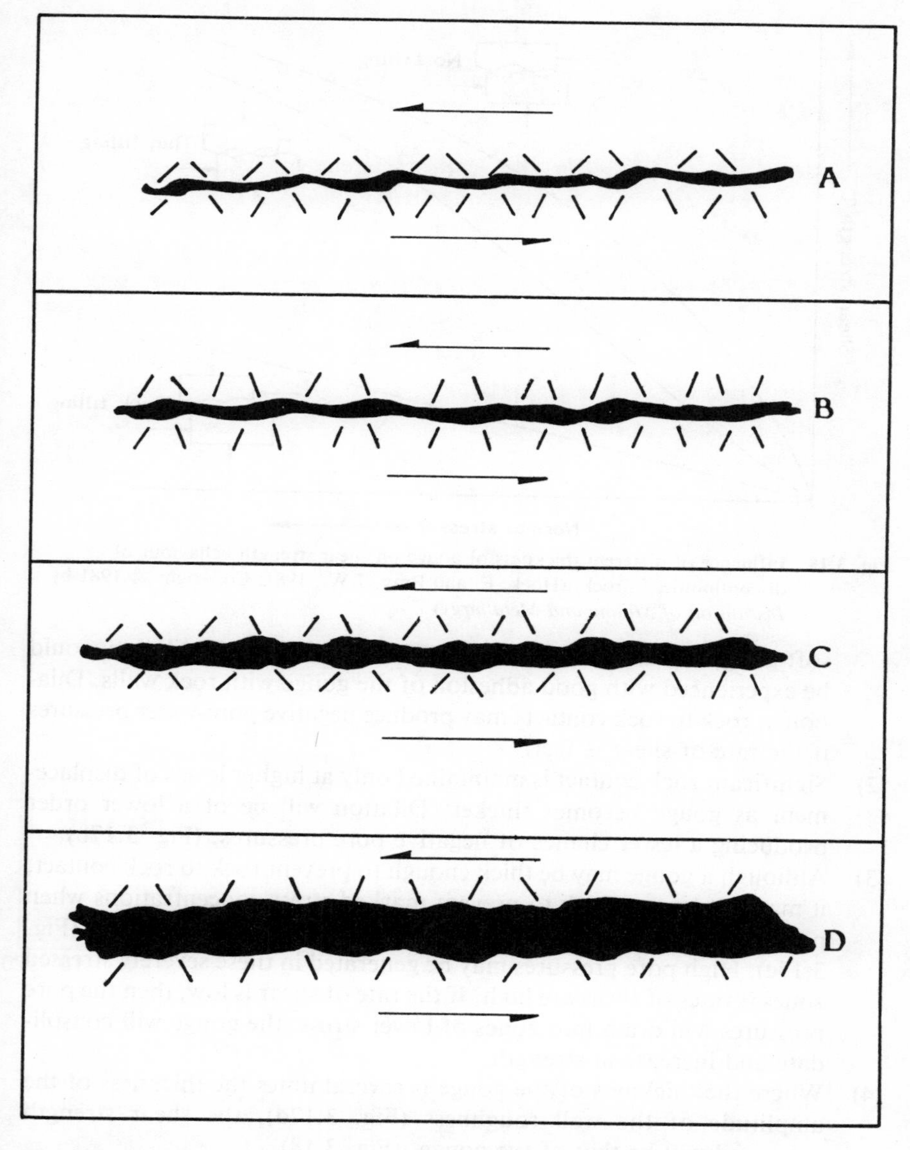

Fig. 3.17 Variations in gouge thickness and the relationship to surface roughness. (Barton, N., 1974; Copyright © 1974 by *Norwegian Geotechnical Institute*)

influence on the shear strength properties of the discontinuity. The range of the influence largely depends on the thickness of the gouge (Barton, 1974).

(1) A very thin gouge (Fig. 3.17a) does not seriously affect good rock to rock contact and strengths would be close to those of the peak strength for infilled discontinuities (Fig. 3.18). At high levels of normal stress,

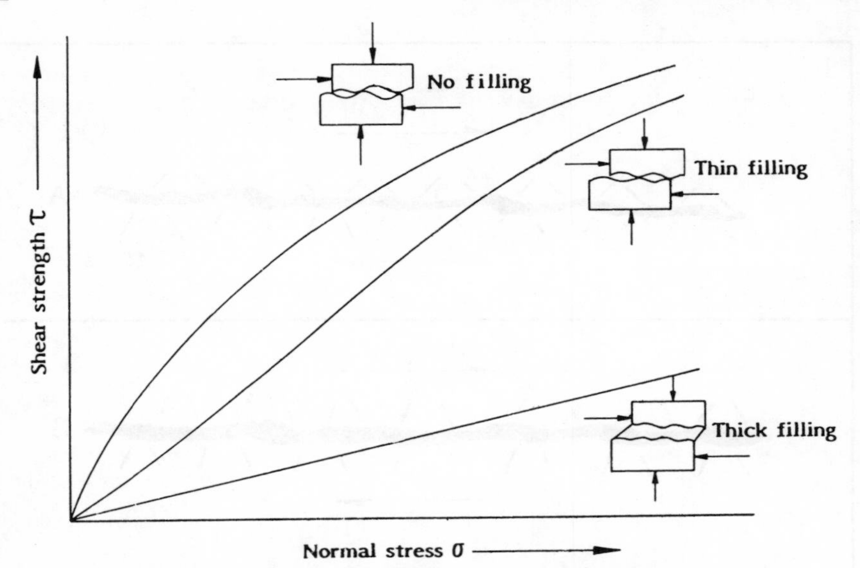

Fig. 3.18 Influence of different thickness of gouge on shear strength behaviour of discontinuities in rock. (Hoek, E. and Bray, J.W., 1981; Copyright © 1981 by *Institution of Mining and Metallurgy*)

soft gouges would be squeezed out. Some reduction in dilation would be experienced with good adhesion of the gouge with rock walls. Dilation at rock to rock contacts may produce negative pore-water pressures if the rate of shear is fast.

(2) Significant rock contact is maintained only at higher levels of displacement as gouge becomes thicker. Dilation will be of a lower order producing a lesser chance of negative pore pressures, (Fig. 3.17b).

(3) Although a gouge may be thick enough to prevent rock to rock contact, it may not be too thick to prevent marked stress concentrations when positive elements of wall roughness are opposite each other (Fig. 3.17c). High pore pressures may be generated in these severely stressed zones if rates of shear are high. If the rate of shear is low, then the pore pressures will drain into zones of lower stress, the gouge will consolidate and increase in strength.

(4) Where the thickness of the gouge is several times the thickness of the amplitude of the wall roughness (Fig. 3.17d), the shear strength measured will be that of the gouge, (Fig. 3.18).

In general, wherever gouge has an influence, shearing behaviour will initially appear to be controlled by the gouge. At higher levels of displacement, the gouge will be punctured and rock to rock contact will produce much higher strengths during dilatant behaviour. At high levels of normal stress, except for extremely thick gouge, displacement of the gouge and shearing through of the joint roughness will be the main influence on strength.

Whatever the thickness of gouge, it should be investigated for mineral composition, grain size and moisture content in order to assist understanding of the patterns of shearing behaviour.

3.4 Residual Stress

In situ stresses are referred to as residual stresses which are made up of two components:
(a) Gravitational Stresses
(b) Latent Stresses

This state exists in a rock mass before the commencement of engineering works.

The natural state of stress that exists at a point within a rock mass is a function of all the previous geologic processes that have acted on the mass. It is impossible to know, with any degree of accuracy what all these events have been. Even if the complete geological history were known, it would not be possible to ascertain the stress state, because the pertinent material properties under long-term loading, and the actual mechanisms of deformation during uplift erosion, etc., are in themselves unknown.

The theory of residual stress was proposed and discussed by Heim in a series of papers (1878–1912), after long work on Trans-Alpine tunnel projects. He believed that there existed a vertical component σ_v probably related to the overburden, but also a horizontal component σ_h of similar magnitude. Heim related these horizontal stresses to major mountain building episodes.

The theory of residual stress employed in soil machanics is dependent much more upon the elastic behaviour of soil under the influence of overburden, where residual stresses result from the resistance to horizontal expansion under vertical compression. As such, the horizontal stress component in this case would be well below the value predicted by Heim.

1 Gravitational stresses

These arise from the superimposed weight of overburden acting upon an element of the rock at depth. The vertical pressure due to a thickness of overburden z of density, γ is given by γ_z. The theoretical solution for the horizontal stresses resulting from such vertical loading is given as:

$$\sigma_2 = \sigma_3 = \frac{\nu}{1 - \nu} \sigma_v$$

where σ_v is the vertical stress
σ_2 and σ_3 are the horizontal stresses
ν is the Poisson's ratio.

Rocks generally have a Poisson's ratio of between 0.2 and 0.3, indicating horizontal stresses between 0.25 and 0.43 times the vertical stress.

By virtue of the nature of rock masses, being cut by joint sets and also often consisting of alternating beds of varied lithological character, the application of a single value for Poisson's ratio is inappropriate and the theoretical ratio of horizontal to vertical stress cannot always apply.

2 Latent stresses

There are several factors to consider in this group:

(a) Stresses induced in past geological time, largely from deep seated stresses. These relate mainly to stresses resulting from tectonic forces. During the main tectonic phase, the resistance to deformation of the rocks builds up until it is in equilibrium with the maximum available tectonc stresses. On attaining this condition, further changes in strain and deformation cease. Rocks in the upper surface of the crust approximate to ideal Bingham bodies, and so the strain energy and associated residual stresses will remain stored in the rock. While there is no further strain, the residual stresses will faithfully represent in quantity and direction the stresses which acted at the end of the tectonic phase of active compression.

(b) Effects of decreasing vertical stresses by erosion of overburden. The adjustment to the new stress level may not be immediate, and it is common in tunnels and underground chambers that values for horizontal stresses exceed those for vertical stresses. The time-lag effect may in fact result in stresses inconsistent with those calculated from thickness and density of overburden; so unpredictable stress conditions may result from erosion.

(c) Latent stresses in rock masses are influenced by topography. The proximity of high ground results in increased stresses in adjacent lower ground.

(d) Low values for vertical stresses may be the result of a bridging effect exerted by strong rocks.

(e) The variable orientation of structural discontinuities will lead to an overall complicated stress distribution in a rock mass. This is further aggravated by variable lithologies, shown by the occurrence of e.g. rock bursts in tunnels and mines in close proximity to faults and areas of rock heterogeneity. Faulting, folding and igneous intrusions all contribute to the production of residual stress.

(f) Swelling of compacted soil materials on absorption of water will tend to lead to the accumulating and release of residual stresses. The stresses are generally stored during the drying-out period as the soil increases in strength and are then released during the next cycle of wetting.

It becomes evident after viewing the complexity of factors initiating residual stresses in rocks, that in practice, there is rarely justification for the assumption that the horizontal stress at a given depth below a horizontal surface is related to the overburden pressure in accordance with elastic theory.

Any one of a number of geologic events could cause the horizontal stress to differ significantly from this value. In an area of active regional subsidence, for example, the centre of the area would be undergoing compressive strain, while the periphery would be undergoing tensile strain. Obviously, the horizontal stresses beneath the centre would be much greater than those at the edges. Similarly, major deep-seated tectonic movements involving convection cells, mountain building, gravity and thrust faulting, would all lead to certain stress states, structural features and boundary conditions which differ greatly from those predicted by elastic theory. Moreover, creep, relaxation of stresses by erosion and weathering, would cause modifications of stresses to the extent that, locally, there would be little resemblance to the initially induced stress field.

3.5 Sheet Joints

This phenomenon, also known as sheeting, is particularly common in igneous intrusions. In areas of, particularly, pronounced topography, sheeting develops parallel or sub-parallel to the surface. In most cases, sheeting reflects topography, and is caused by a combination of factors, including:

(a) differential expansion of minerals on weathering;
(b) removal of load of superincumbent rock by erosion;
(c) seasonal temperature variations near the rock surface.

Sheeting commonly occurs during quarrying and excavations. They form suddenly and with a low pitched report.

In cuttings and cliff areas, extra sets of joints are often found. These are believed to result from gravitational sliding of sheets. Oblique joints probably reflect obstructions to the free sliding of the sheets, or are strongly impressed incipient joints opening at the time of available expansion. Such "contour joints" are not restricted to igneous rocks.

New sets of joints do not always develop as a result of stress relief. If pre-existing discontinuities are sympathetically aligned with respect to the excavation, then it is much more likely that the stress itself will be manifested as opening of these discontinuities rather than the formation of new ones.

CHAPTER 4: WEATHERING, EROSION, TRANSPORTATION AND DEPOSITION

4.1 Weathering and its Significance in Geotechnical Engineering

Weathering may be defined as

> "that process of alteration of rocks occurring under the direct influence of the hydrosphere and atmosphere".

Rock weathering is important in geotechnical engineering since it concerns the behaviour of materials used as embankment fill, concrete, roads, or buildings. It is also concerned with the behaviour of weathered materials in rock structures, e.g. in the foundations of the St Francis and Malpasset dams.

1 Weathering processes

Weathering of a rock by physical breakdown without considerable change in mineralogy is called disintegration, and the residual soil is an accumulation of mineral and rock fragments virtually unchanged from the original rock. This type of weathering is found particularly in arid and cold climates.

Chemical alteration of the minerals in a rock is called decomposition, and significantly affects most rock forming minerals except quartz. The greater the proportion of susceptible minerals in a rock, the more obvious is the change.

2 Weathering and weatherability

Many rocks were formed at high temperatures and pressures and much of weathering is the changing of rocks to a new state of stability at lower temperatures and pressures in the presence of air and water. These processes virtually always mean a reduction in the quality of the geological materials in terms of their engineering properties.

Weathering along faults and joints has been found at considerable depths during site investigations, e.g. Keiwa river HEP project, Victoria, Australia where it was discovered in gneisses at a depth of 350 metres.

Weathering occurs over very long periods of time, although durability of rocks may be in question over much shorter periods, when rocks e.g. in a slope, or used as construction materials, continue to weather. This short-term susceptibility to alteration is defined as weatherability of a material. It is of considerable importance in civil engineering, e.g. in the deterioration of rocks and soils in freshly excavated slopes, and of rocks to be used for aggregates. Weatherability here is often associated with the behaviour of rocks when they are in contact with water. This is particularly important in mudrocks and weathered rocks and is called slaking.

The main results of weathering are:

1. Production of more stable minerals.
2. Breakdown of geological materials from a massive to a clastic or plastic state.
3. Changes in volume, density, particle size distribution, surface area, porosity, permeability, compressibility and strength.
4. The formation of new minerals, aggregates and solutions.
5. Changes in the durability of original minerals.
6. The redistribution of minerals and salts.
7. The preparation of rock surfaces for erosion.
8. The formation of new land surfaces and deposits.

Rate of weathering depends on:

1. Environmental factors
   (a) climate
   (b) hydrology
   (c) biology
   (d) time

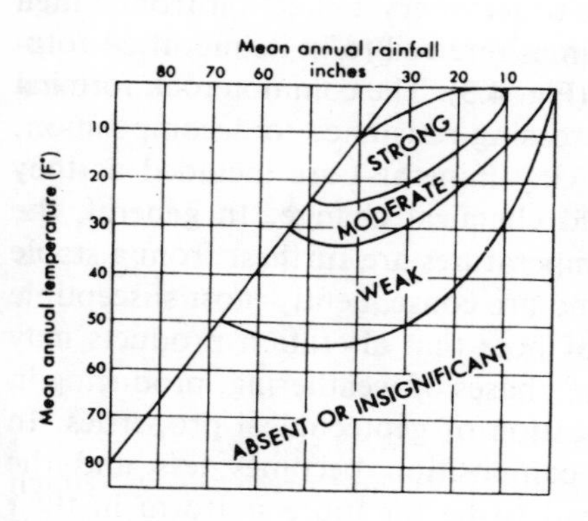

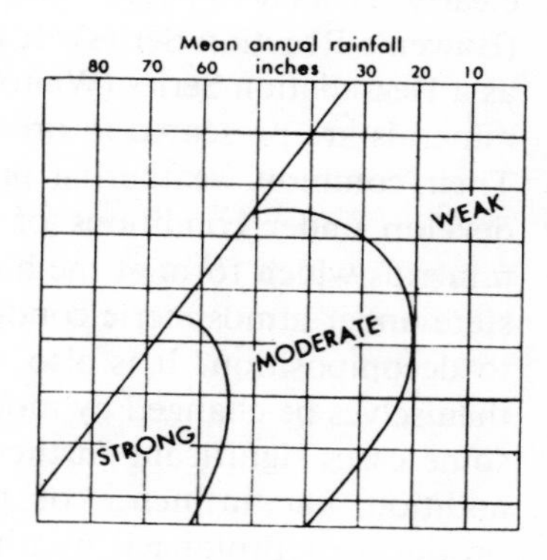

Fig. 4.1 Influence of mean annual temperature and mean annual rainfall on
a. disintegration weathering and
b. decompositional weathering
(Peltier, L.C., 1950; Copyright © 1950 by *Association of American Geographers*)

2. Parent Material
   (a) mineralogy
   (b) proportions of different minerals (mode)
   (c) induration
   (d) permeability
   (e) existing grade of weathering
3. Structure
   (a) discontinuity spacing
   (b) openness of discontinuities

The climatic factors which are usually emphasized in evaluations of weathering characteristics are rainfall and temperature. By using a graph of mean annual temperature against mean annual rainfall, (Peltier, 1950) clear zones are defined which indicate the relative importance of decomposition and/or disintegration under different climatic regimes (Fig. 4.1). The effect of biologial influences, e.g. root wedging, provision of organic acids, etc. is implicit in the classification based on climatic regimes. The Russians' view of such a zonation is similar (Strakhov, 1967) and works on an essentially latitudinal basis, (Fig. 4.2).

Such classifications are through necessity generalized, and other factors tend to cause distortion of these zones. The assumptions that higher temperatures and amounts of rainfall accelerate chemical weathering is generally reasonable, as is the converse. This does not mean though that local differences do not exist.

Mineralogy has a major influence on the properties of weathering in that common rock forming minerals have considerably different durabilities. It is the changes to these minerals which constitute the most important factor in the make-up of a rock relative to its susceptibility to weathering. The work carried out on relative susceptibility of minerals to decomposition clearly shows it to be the reverse of the order of crystallization from a melt (Bowen's Reaction Series). It is sometimes referred to in its modified form as a Degradation Series (Walton 1971) (Fig. 4.3). The common rock forming minerals are presented in order of increasing resistance to decomposition. Their common weathering products (clay minerals) are included as they develop under conditions of increased chemical change. In general, the minerals which form at the highest temperatures are furthest from a stable state under atmospheric conditions, and are consequently most susceptible to decomposition. It is also worthy of note that alteration products may themselves be changed by more intense phases of weathering, producing in some cases significant further modification of geotechnical properties. In addition, the influence of original composition becomes less and the products of prolonged weathering tend to be far more uniform in their characteristics than the rocks from which they were formed.

Grain size of an original rock has a distinct bearing on rates of weathering. Coarse-grained rocks tend to have a greater porosity through their larger pore sizes. This enables a greater absorption of water to be achieved,

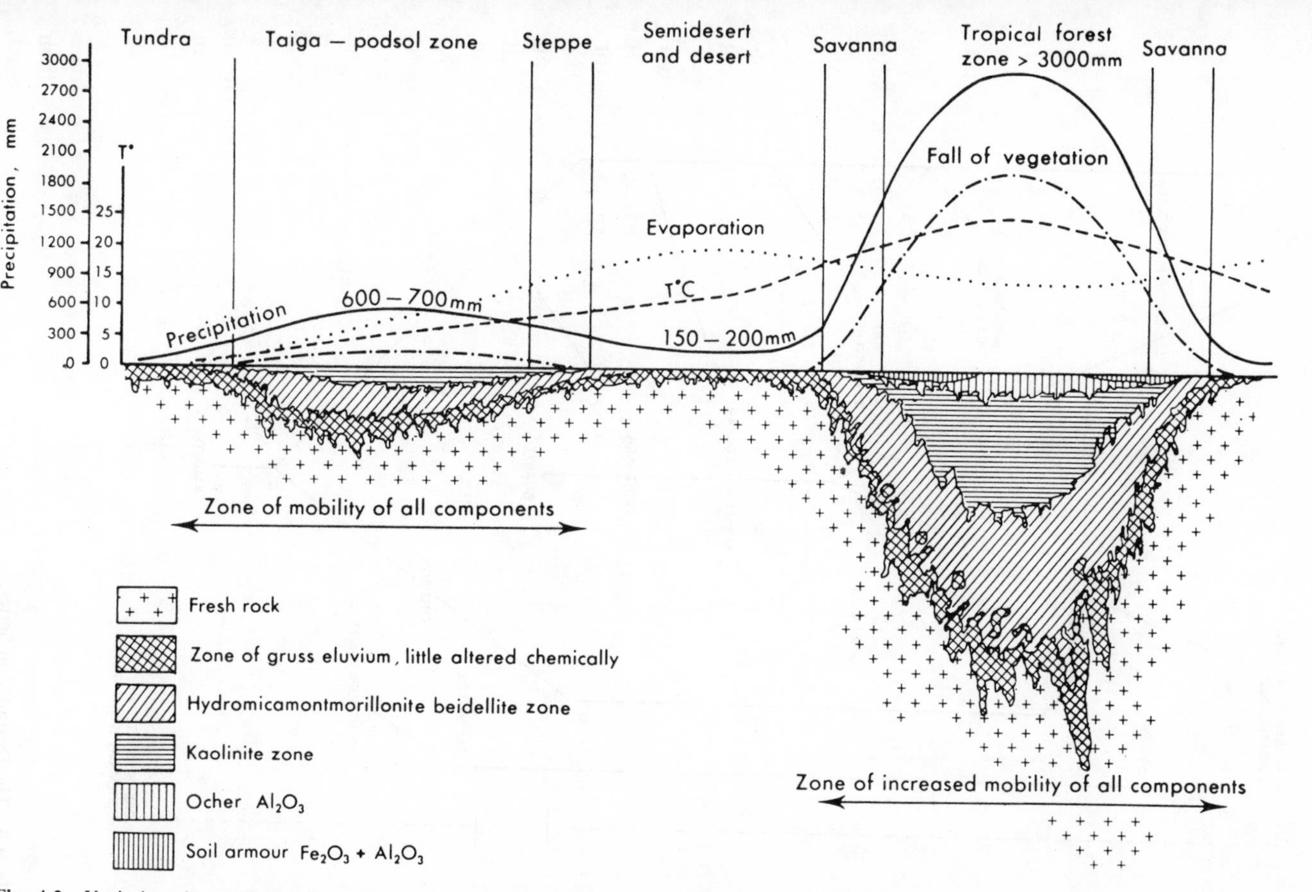

Fig. 4.2 Variations in weathering intensity with latitude. (Strakhov, N.M., 1967; Copyright © 1969 by Longman, London)

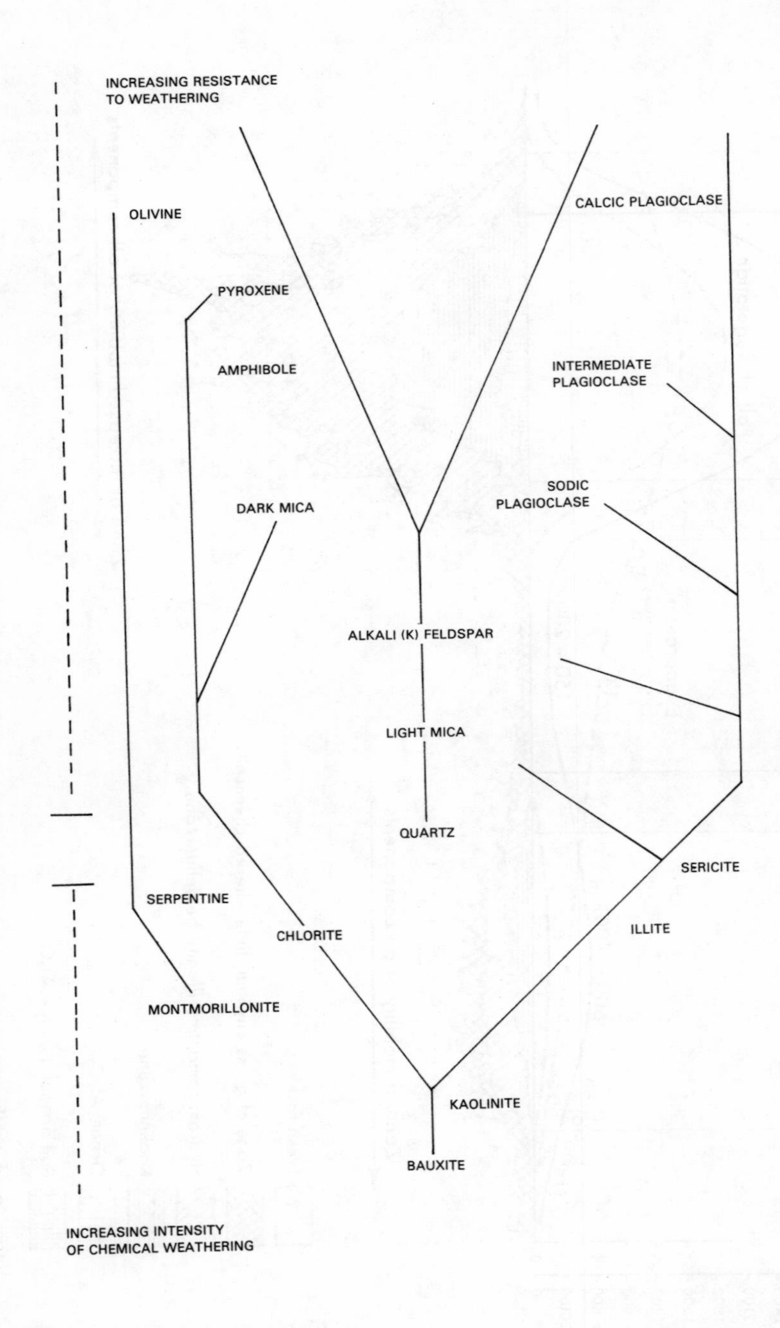

Fig. 4.3 The Degradation Series.

and such rocks tend to weather relatively quickly. The rocks with a crystalline texture especially igneous rocks, have low porosities and intercrystal permeabilities. In the mass, water ingress is along discontinuities. This helps to account for the fact that corestones frequently develop as a result of the weathering of igneous rocks, but rarely of clastic rocks. However, differences in weathering rates due solely to grain size contrasts have been

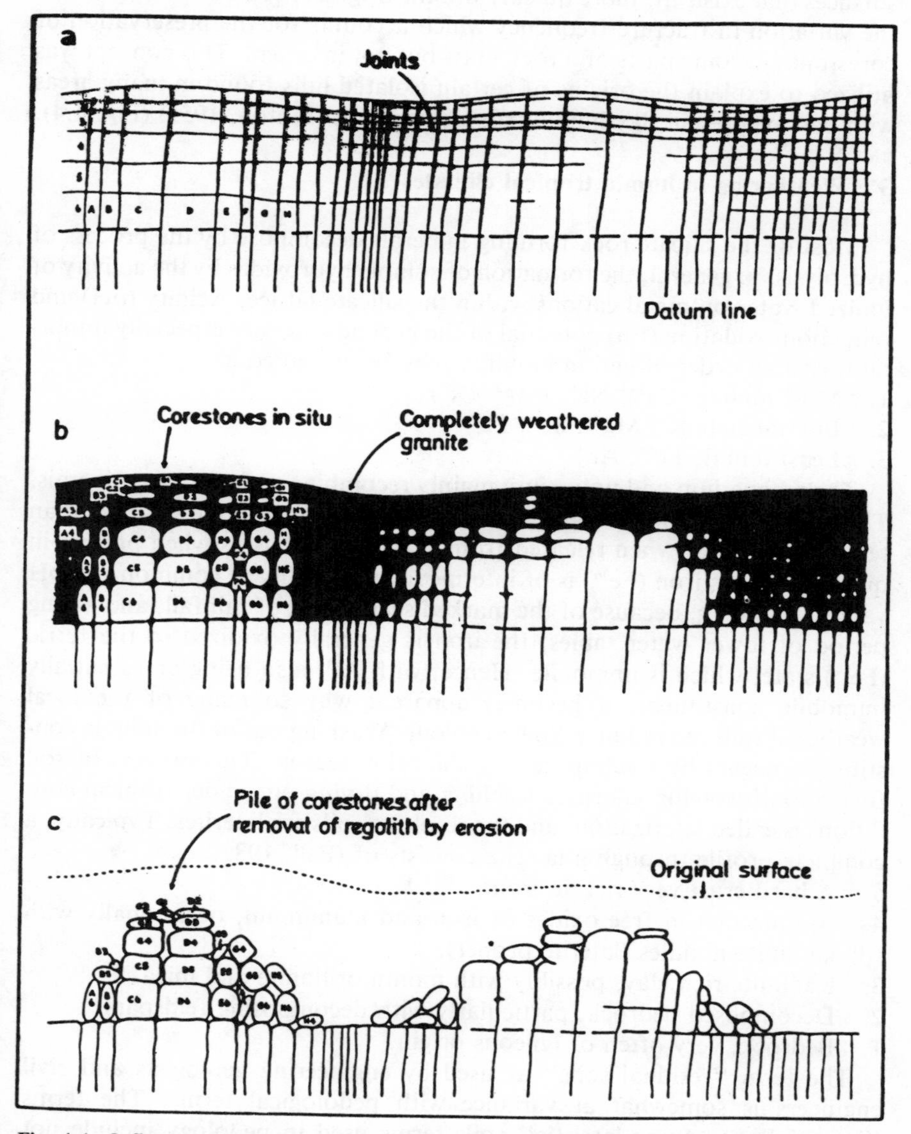

Fig. 4.4 Influence of fracture frequency on the intensity of decompositional weathering. (Ollier, C.D., 1969; Copyright © 1969 by Longman, London)

encountered where granites in a completely weathered state are found intruded by finer grained dykes which are still relatively fresh.

The structure of the rock mass is of great importance to the rate and pattern of weathering which develops. The discontinuities which cut a rock mass afford routes for the penetration of weathering agencies. Weathering processes attack the rock from the discontinuity surfaces, so the more surfaces that exist, the more quickly will the original rock be consumed. It is the variation in fracture frequency which accounts for the preservation of corestones in some parts of a rock mass but not in others. This concept was utilized to explain the origins of certain isolated hills found in many areas with past or present day humid tropical climates (Linton, 1955) (Fig. 4.4).

3 Weathering in humid tropical climates

Most of the silicate rock forming minerals decompose by the process of hydrolysis, in general, the formation of soluble hydroxides by the activity of ionized water on metal cations within the silicate lattice. Acidity (pH) and reduction-oxidation (Eh) potential of the groundwater are especially important, and an order of cation mobility may be presented as:

1. Most mobile, Ca^{2+}, Na^{+}, (Mg^{2+})(K^{+})
2. Intermediate K^{+}, Mg^{2+}, Si^{4+}, Fe^{2+}
3. Least mobile Fe^{3+}, Al^{3+}

The magnesium and potassium mainly recombine to form clay minerals. The silicon has a low but significant mobility, and its solubility is more than ten times as great when released from silicate minerals as when present in quartz. Ferrous iron (Fe^{2+}) is of intermediate solubility at common soil pH values. However, because of the marked seasonality of rainfall, and a long period of lower water tables, the iron is generally oxidized to the ferric (Fe^{3+}) state, which is immobile. Hence, with Al^{3+} also being an essentially immobile constituent, it becomes apparent why so many of the local weathered soils are red or yellow in colour. Washing out of the soluble constituents occurs by leaching during the rainy season. This process of soil formation involving alternate leaching and drying out under tropical conditions is called laterization, and the resulting soils are laterites. Typically, a complete profile through a laterite consists of (Bell, 1981):

5. A hard crust very rich in iron.
4. A zone rich in free oxides of iron and aluminium, occasionally with kaolinite nodules (laterite proper).
3. Kaolinite-rich clay, possibly with montmorillonite and mica.
2. Decomposed bedrock, particularly with decomposing feldspars.
1. Bedrock, very often of igneous origin.

The term "residual soils" as used by engineering geologists and civil engineers is somewhat at variance with pedological terms. The terms "lateritic" and "non-lateritic" soils terms used in pedology include not only Grade VI (see Table 4.4) weathered material, but also some Grade V, as indicated by the description of some lower units in such soils as retaining the original bedrock structure. For the most part, Grade VI materials only will be considered as constituting residual soil. In both igneous and sedimentary rocks, original textures and fabrics are usually discernible in the Grade V materials, even though most of the original rock forming minerals except for quartz, have been altered, usually to clay minerals.

Residual soils develop under the influences of soil (pedological sense) forming processes. These are much more complex than the processes decomposing the rock and result in considerable variation within a residual soil profile. Most of the work undertaken on weathered materials in humid tropical environments has been on those overlying igneous rock. This has concentrated on the gradation of the weathering product from the bedrock surface or, from the applied point of view, the variation in the amount and distribution of corestones with depth. On sedimentary materials, the problem of corestones does not often appear to arise, probably because of the inherent primary porosity of the materials. There is however considerable variability in the weathering profile on both igneous and sedimentary successions as indicated by the results from in situ penetration testing during site investigations and from investigations of slope failures.

The main emphasis in most studies has been on chemical and mineralogical composition notably types of clay minerals formed, and also on zones of iron oxide redeposition. A description of the physical variability within the profile will be outlined here as being of more importance in geotechnical engineering. Zones of depletion and enrichment of, e.g. fine-grained particles, may exist. A variety of factors must be considered, and for any parent material, these may produce residual soils of quite different character. It is not therefore possible to consider "typical" grading curves, but some samples from a residual soil overlying a mudrock from western Singapore showed uneven particles size distributions strongly indicating depletion of part of the silt fraction producing a gap graded soil (Fig. 4.5). Samples of residual soils were taken within 100 mm of the junction of six different sedimentary bedrock types, and a summary of grading and plasticity characteristics is shown in Table 4.1. Three examples are shown of profiles through residual soils overlying different bedrock types (Table 4.2). Overall, the finer the parent rock, the finer the resulting soil, although the formation of nodules of iron rich clay is quite common, and will extend the particle size distribution curve. The significance of bedrock type on the composition of residual soil tends to be greater with sedimentary rocks, especially mixed sedimentary successions, than with igneous (Nossin and Levelt, 1967) or metamorphic rocks (Sinclair, 1980) (cf Figs. 4.6 and 4.7). In each case however, the range of particle sizes present is a reflection of the "layering" which is apparent within the residual soils.

Plasticity characteristics may also give some guidance to the nature and trend of changes in residual soils. The clay minerals present will have an

Table 4.1 Characteristics of residual soils near to the bedrock contact.

	Bedrock Type	Particle Size Distribution				Plasticity Characteristics		
		% Gravel	% Sand	% Silt	% Clay	W_L %	W_P %	PI %
A	Cleaved dark red silty, slightly clayey MUDROCK	8	21	39	32	46	24	22
B	White and red mottled silty CLAY	8	27	33	32	44	24	20
C	Cream to pale brown silty CLAY	6	28	40	26	36	28	8
D	Purple cemented silty clayey MUDROCK	3	22	46	29	42	18	24
E	White and yellow mottled sandy, silty MUDROCK	3	19	43	35	44	23	21
F	Brown sandy, clayey SILTSTONE	4	11	51	34	44	25	19

Table 4.2 Residual soil profiles.

Distance from contact (cm)	% Gravel	% Sand	% Silt	% Clay	W_L %	W_p%	PI %
Profile over pale brown sandy silty CLAY							
40	0	22	35	43	52	27	25
80	0	27	35	38	52	28	24
120	0	17	40	43	52	29	23
140	0	20	36	44	57	28	29
Profile over yellow with red mottling, fissured silty CLAY							
20	0	13	13	74	67	36	31
40	0	12	13	75	67	35	32
60	0	10	15	75	67	33	34
80	5	20	9	66	69	34	35
100	14	21	5	60	70	35	35
120	4	21	10	65	68	31	37
140	10	25	7	58	64	23	41
160	0	16	10	74	68	35	33
180	0	15	10	75	69	35	34
Profile over purplish-red slightly clayey silty fine and medium feldspathic SANDSTONE							
10	1	51	21	27	49	27	22
20	5	44	17	34	57	33	24
30	9	60	8	23	63	29	34
40	20	54	6	20	56	26	30
50	18	62	6	14	58	25	33

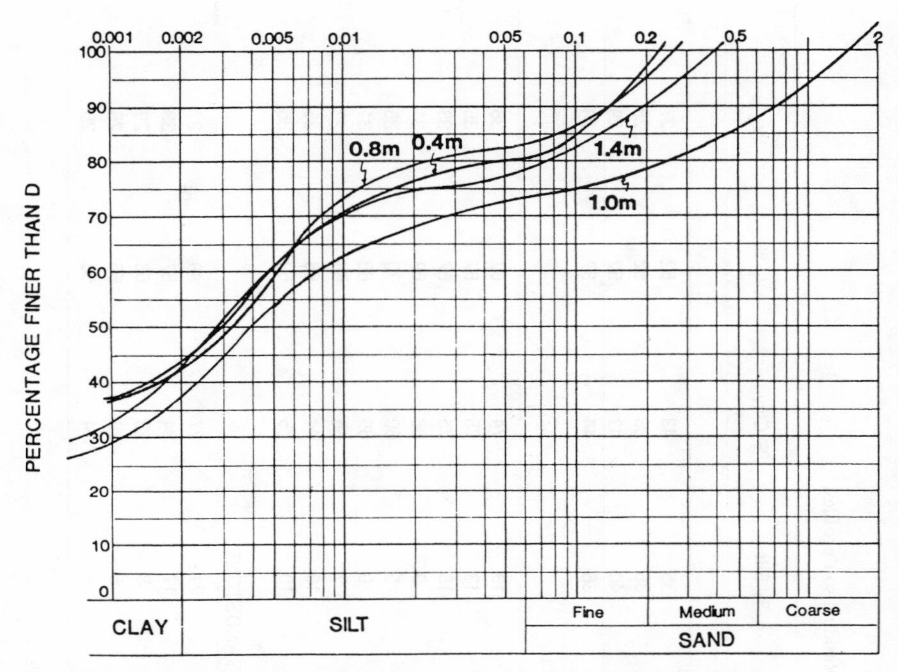

Fig. 4.5 Particle size distribution curves for residual soils overlying clastic sedimentary rocks in part of western Singapore.

influence on the plasticity characteristics measured, although decreases in plasticity would also result from incompletely weathered original minerals remaining in the soil. Plasticity tends to increase with depth, confirming a tendency for a downward migration of fines in lateritic soils. A summary of the plasticity characteristics of some South-East Asian lateritic soils is given in Table 4.3 (Nixon and Skipp, 1957a; West and Dumbleton, 1970). It is well known that when dealing with tropical residual soils, different amounts of work imposed on liquid limit samples leads to variations in liquid limit values (Moh and Mazhar, 1969). The normal trend is for additional work to break up soil aggregations, thereby increasing the effective area of clay platelets available to take on adsorbed water. This causes a gradual increase in the liquid limit·with time. Exact comparability of results from source to source is therefore unlikely.

The results of the plasticity tests shown for nine sedimentary rocks in Singapore (Fig. 4.8) reveal quite a wide range of values. The points generally fall very close to the A line, with a distinct majority plotting just above.

The results of the classification tests illustrate the non-homogeneous nature of residual soils. Single values describing particle size or plasticity characteristics quoted in the literature are not very meaningful.

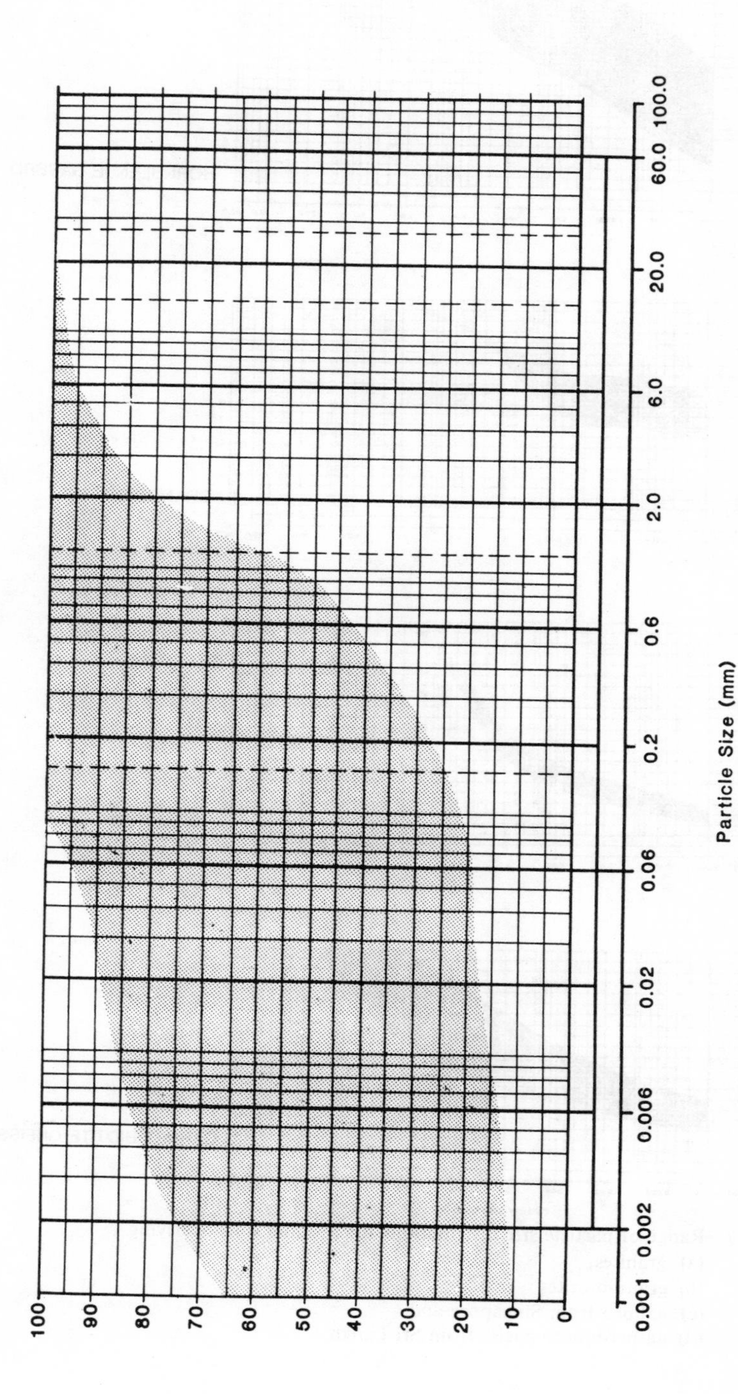

Fig. 4.6 Range of particle size distribution of residual soils from a mixed sandstone-mudrock succession of the Jurong Formation, Singapore.

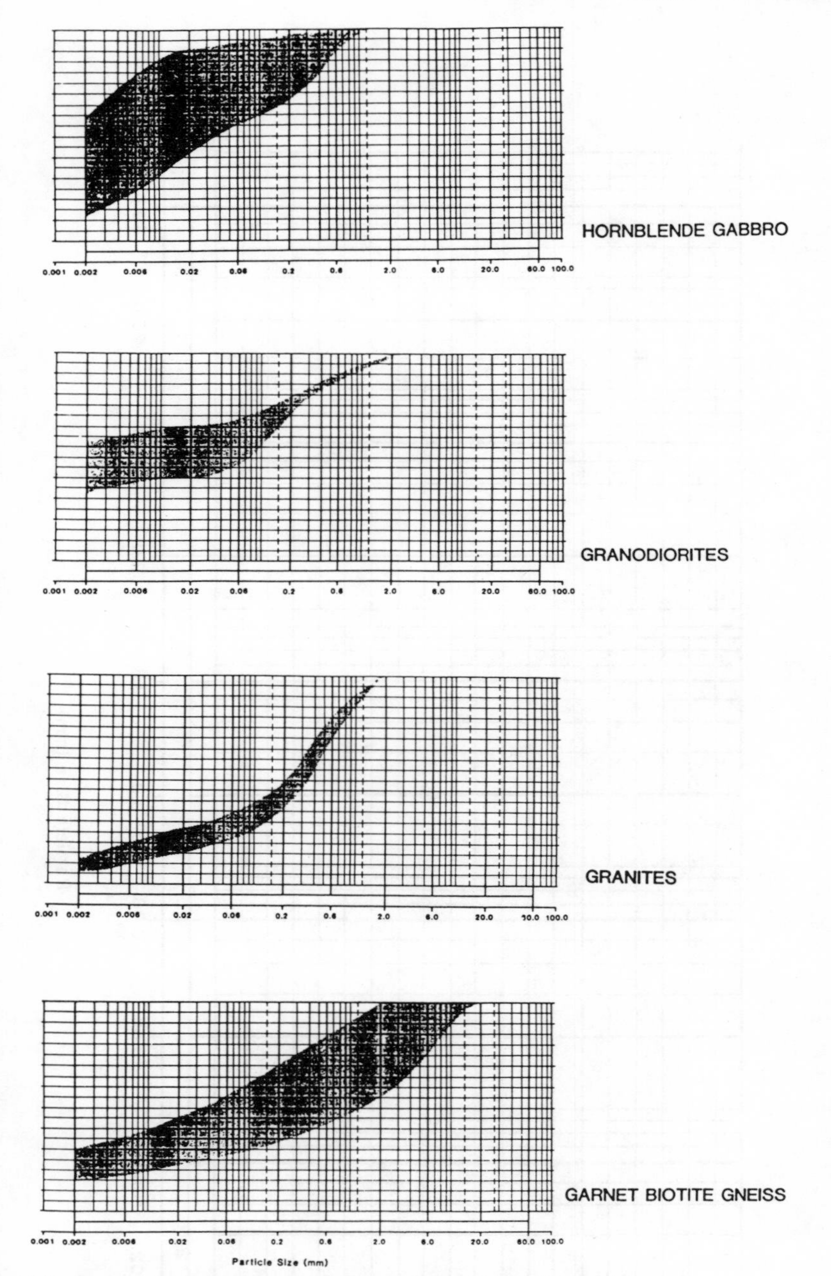

Fig. 4.7 Range of particle size distributions for residual soils overlying
(a) granites,
(b) grano-diorites
(c) gabbro from Singapore and
(d) garnet-biotite gneiss from Sri Lanka

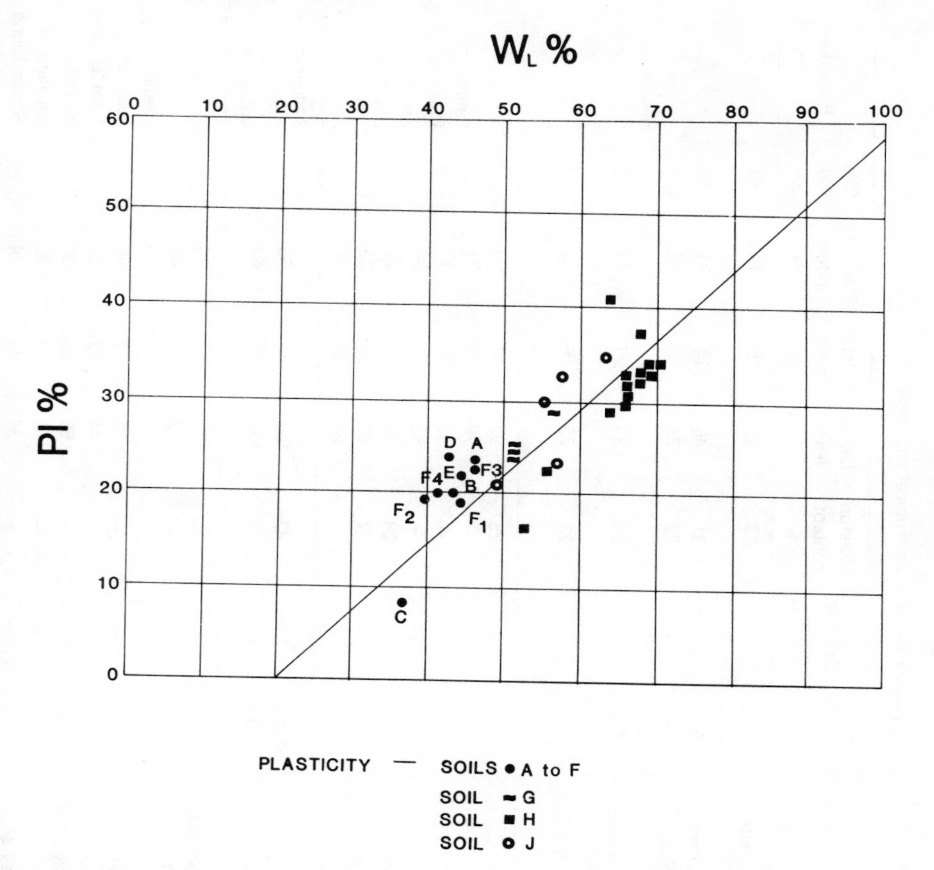

Fig. 4.8 Range of plasticity characteristics for residual soils from nine sedimentary bedrocks from Singapore.

Some lateritic soils develop hard surface crusts which have high strength and are incompressible. The layers beneath may have less good bearing capacities and true lateritic soils frequently show a decrease in shear strength with depth. A series of undrained triaxial tests from Sri Lanka showed such a decrease in strength from 90 kN/m² to 25 kN/m² from the surface to a depth of 6.0 m (Nixon and Skipp, 1957b). Angles of shearing resistance tend to be fairly consistent and between 25° and 35°. The higher values correspond with the coarser grained soils. A most important feature is that most tropical residual soils have cohesive as well as frictional strength and will often stand in vertical cuts up to about 6.0 metres deep. As soon as they become saturated, however, the cohesion tends towards zero and both completely weathered rocks and residual soils are prone to complete disintegration. Some lateritic soils are extensively water bearing and pumping may be required during excavation.

Table 4.3 Series of results of classification tests on tropical red clays.

Location	Sample description	Depth m	Specific gravity	Moisture content: %	Liquid limit: %	Plastic limit: %	Plasticity index: %	Clay (2) %	Remarks
Malacca, Malaya	Mottled brown and white soft silt, with fine sand and clay	8	—	33	29	15	14	12	Alluvial
	Soft light grey clay with red iron stains	9	—	31	45	14	31	38	
	Chocolate-coloured soft to firm micaceous silt	9	—	22	28	16	12	16	
	Soft mottled grey and yellow micaceous silt and clay	6.5	—	25	33	20	13	16	
	Pink and brown clay and silt with medium sand and gravel	8	—	23	25	16	9	14	
Pahang, Malaya	Chocolate inorganic clay	—	—	25	99	44	55	—	Pahang Volcanic Series (residual)
Central Johore, Malaya	Light purple inorganic clay	—	—	—	70	37	33	—	
	Red inorganic clay	—	—	—	35	22	13	—	
	Dark grey inorganic clay	—	—	—	53	27	26	—	
	Light brownish-grey inorganic clay	—	—	28	74	35	39	—	
	Ochre inorganic clay	—	—	27	67	34	33	—	
	Red inorganic clay	—	—	—	87	42	45	—	
Tylilitan Aerodrome, Jakarta, Java	Red clay	0–0.09	2.85	42	100	45	55	—	? alluvium from volcanics
	" "	0–0.09	2.72	—	102	40	62	—	
Paya Lebar Airport, Singapore	Red and yellow brown clayey silty sand	—	—	—	57	22	35	—	Results selected from 33 tests on older alluvium of S'pore Island
	Mottled red and yellow clayey silt, with traces of fine sand	—	—	—	72	32	40	—	
	Brown clayey coarse sandy gravel	—	—	—	75	27	48	—	
	Yellowish clayey sand	—	—	—	44	22	22	—	
	Mottled red and yellow clayey silty sand	—	—	—	68	33	35	—	
	Mottled red and yellow silty clay	—	—	—	68	20	48	—	
	Purple and yellowish-brown clayey silt	—	—	—	34	20	14	—	
	Mottled red and yellow silty clay	—	—	—	63	34	29		

Johore 1.8 to 2.0 m/y, 27 to 28°C	Reddish brown clay	1.5	92	44	48	69	Overlying
	Laterite gravel with brownish-red clay	2.4	105	46	59	18	basalt
Kuantan-Sungei Lembing Road, near crossing over Sungei Pinang, Pahang 3.1 to 3.6 m/y, 25 to 26°C	Laterite gravel with dark reddish-brown friable clay	0.6	52	33	19	17	
	Laterite gravel with dark reddish-brown friable clay	1.8	46	31	15	24	Overlying
	Dark brownish-red friable clay with laterite gravel	3.1	—	—	—	9	basalt
	Dark brownish-red friable clay with a little laterite gravel	7.6	—	—	—	19	
Muar-Segamat Road, mile 31, Johore 1.8 to 2.0 m/y, 27°C	Brown clay with fine mica and a little angular quartz	0.6	43	23	20	37	Overlying
	Brown silty clay with gravel	1.8	43	22	21	26	granite
	Brown sandy clay with fine mica and quartz	2.3	42	21	21	33	
Tampin-Gemas Road, mile 41½ from Seremban, Negri Sembilan 1.8 to 2.0 m/y, 25 to 26°C	Light brown silty clay with some angular quartz	0.8	84	32	52	49	
	Light pinkish-brown silty clay with some angular quartz	2.1	106	34	72	54	Overlying granite
	Gravel, sand and pink clay	3.7	107	33	74	34	
Kampong Jeram, Alor Star-Jitra Rd, mile 17, Kedah 2.0 to 2.3 m/y, 27°C	Laterite gravel with brown clay	0.6	—	—	—	29	Overlying
	Very light brown silty clay	1.5	64	33	31	43	sedimentary
	Black silt	6.1	51	29	22	37	rocks
Muar-Pagoh Road, mile 15, Johore 1.8 to 2.0 m/y, 27 to 28°C	Laterite gravel with some light reddish-brown clay	1.4	96	37	59	14	Overlying sedimentary
	Very light yellowish-brown silty clay	5.6	84	50	34	47	rocks

Table 4.3 (cont'd)

Table 4.3 Series of results of classification tests on tropical red clays. (cont'd)

Location	Sample description	Depth m	Specific gravity	Moisture content: %	Liquid limit: %	Plastic limit: %	Plasticity index: %	Clay (2) %	Remarks
Batu Tiga Industrial									
Site, near Klang,	Brown clay	0.6			41	20	21	41	Overlying
Selangor	Light reddish-brown clay	1.8			46	23	23	46	sedimentary
2.3 to 2.5 m/y,	Light brownish-red clay	3.4			50	22	28	46	rocks
26 to 27°C									
Nanyang Tech.	Reddish-yellow sandy clay	0.4		31	57	28	29	43	Overlying
Institute	Reddish-yellow sandy clay	0.8		35	52	29	23	43	sedimentary
Singapore	Reddish-yellow sandy clayey gravel	1.0		26	52	28	24	37	rocks
	Reddish-yellow sandy clay	1.4		35	52	27	25	44	

Prolonged periods of leaching (downward movement of water) during wet periods may cause the breakdown of aggregations (lateritic nodules) into their constituent clay particles. An increase in the liquid limit is one result of this. An increase in compressibility perhaps by 50% or more, and a decrease in the coefficient of consolidation (C_v) by up to about 20% may also occur (Ola, 1978).

4 Tests to assess weatherability

Weatherability should be studied for the full life of an engineering structure to determine how much weathering will occur during this period. Some "speeding-up" process is usually adopted, although this may be a questionable practice. Griggs (1936) heated granites rapidly by electric heater and cooled them in a stream of dry air to produce a temperature variation of 110 °C. He simulated 244 years of diurnal temperature change and found no detectable change in the granite surface. However, by cooling the rock with tapwater, he found that the equivalent of 2 years of weathering caused loss of polish, surface cracking and the beginning of surface disintegration. He concluded that chemical weathering was more important than insolation weathering.

Durability of rocks must be checked to ensure that no serious deterioration occurs whilst in place in an engineering structure. This resistance to short-term weathering is especially important for mudrocks and partly weathered materials and the slake durability test is used to check competency (Franklin and Chandra, 1972). The test involves gentle agitation of fragments of rock in water in a sieve, and a measure of the fines produced in a particular time. It has been used to assess suitability of roadstones, and riprap.

4.2 Engineering Classification of Weathered Rock

Unqualified rock names may be misleading if mechanical properties are implied from them. The use of weathering grades, or even better, index property tests are favoured to indicate likely physical performance of a weathered rock. The rock should be described by its name, weathering grade, fracture spacing and strength indices (Anon, 1970; Anon, 1977).

A. Weathering grade classification. (See Table 4.4.)

B. Fracture classification.

Weathering is often accompanied by an increase in the intensity of fracturing, i.e. a decrease in fracture spacing, and fracture surfaces often exhibit decay as a result of weathering. The depth of weathering penetration largely depends on rock type, and especially permeability, e.g. porous sandstones may weather throughout, whereas relatively non-pervious igneous rocks weather along fractures and leave intact kernels of rock in the middle

Table 4.4 Engineering grade classification of weathered rock.

Grade	Degree of decomposition	field recognition: Soils (i.e. soft rocks)	field recognition: Rocks (i.e. hard rocks)	Engineering properties
VI	Soil	The original soil is completely changed to one of new structure and composition in harmony with existing ground surface conditions.	The rock is discoloured and is completely changed to a soil in which the original fabric of the rock is completely destroyed. There is a large volume change.	Unsuitable for important foundations. Unstable on slopes when vegetation cover is destroyed, and may erode easily unless a hard cap present. Requires selection before use as fill.
V	Completely weathered	The soil is discoloured and altered with no trace of original structures.	The rock is discoloured and is changed to a soil, but the original fabric is mainly preserved. The properties of the soil depend in part on the nature of the parent rock.	Can be excavated by hand or ripping without use of explosives. Unsuitable for foundations of concrete dams or large structures. May be suitable for foundations of earth dams and for fill. Unstable in high cuttings at steep angles. New joint patterns may have formed. Requires erosion protection.
IV	Highly weathered	The soil is mainly altered with occasional small relic blocks of original soil. Little or no trace of original structures.	The rock is discoloured; discontinuities may be open and have discoloured surfaces and the original fabric of the rock near the discontinuities is altered; alteration penetrates deeply inwards, but corestones are still present.	Similar to grade V. Unlikely to be suitable for foundations of concrete dams. Erratic presence of boulders makes it an unreliable foundation for large structures.
III	Moderately weathered*	The soil is composed of large discoloured relic blocks of original soil separated by altered material. Alteration penetrates inwards from the surfaces of discontinuities.	The rock is discoloured; discontinuities may be open and surfaces will have greater discolouration with the alteration penetrating inwards; the intact rock is noticeably weaker, as determined in the field, than the	Excavated with difficulty without use of explosives. Mostly crushes under bulldozer tracks. Suitable for foundations of small concrete structures and rockfill dams. May be suitable for semipervious fill. Stability in cuttings depends on

	weathered	angular blocks of fresh soil, which may or may not be discoloured. Some altered material starting to penetrate inwards from discontinuities separating blocks.	discoloured; particularly adjacent to discontinuities which may be open and have slightly discoloured surfaces; the intact rock is not noticeably weaker than the fresh rock.	Suitable for concrete dam foundations. Highly permeable through open joints. Often more permeable than the zone above or below. Questionable as concrete aggregate.
I	fresh rock	The parent soil shows no discolouration, loss of strength or other effects due to weathering.	The parent rock shows no discolouration, loss of strength or any other effects due to weathering.	Staining indicates water percolation along joints; individual pieces may be loosened by blasting or stress relief and support may be required in tunnels and shafts.

*The ratio of original soil or rock to altered material should be estimated where possible.

of each block. Rocks of the sandstone type show thick development of Grade IV weathering; igneous rocks, of Grade III.

The gradation of weathering throughout a rock mass, particularly one composed of igneous rock, would perhaps be like the mass portrayed in Fig. 4.9. However, in some areas of humid tropical weathering on low lying areas with low angled slopes it is often the case that corestones do not survive, and that Grade V or Grade VI material rests directly on a very thin zone of Grade III and Grade II rock, in turn overlying fresh rock (Fig. 4.10).

The strength of rock joints largely depends on their surface roughness. If these are heavily weathered, they will shear through, rather than dilating by over-riding. If shearing occurs largely in weathered material on a joint, the stronger remnants will contribute little to the strength of the rock mass.

The fractures in a rock mass are important from the point of view of, orientation, tightness, roughness, infilling material or weathered lining. The spacing is important since this determines freedom for displacements or fluid movements within the fractured mass.

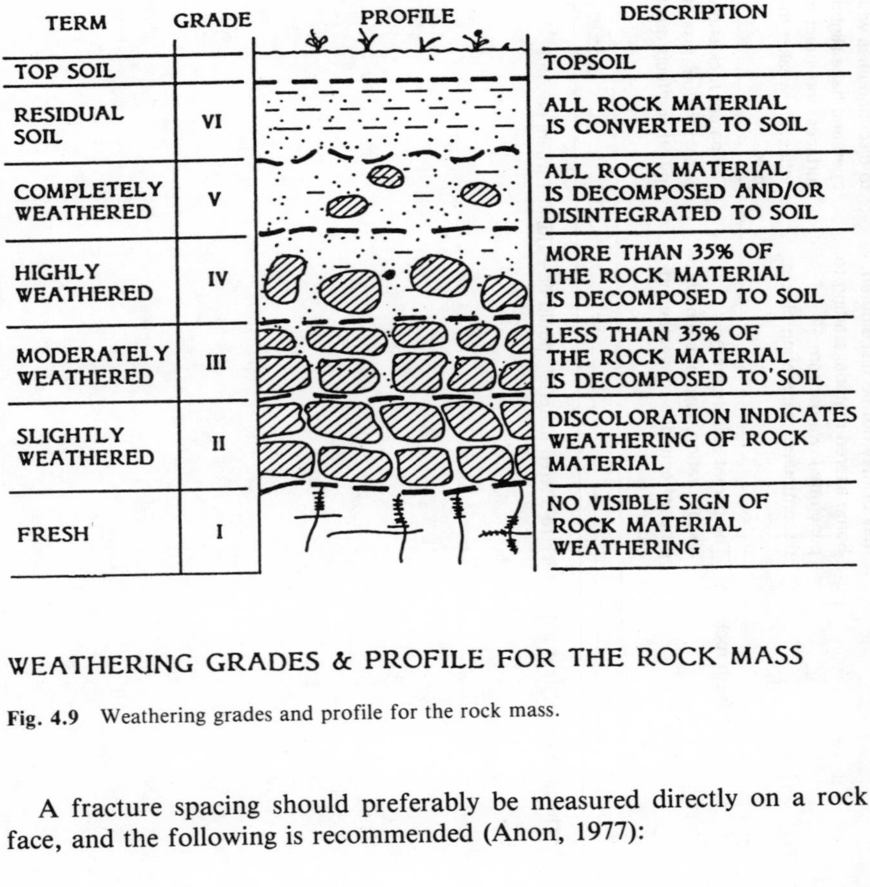

TERM	GRADE	PROFILE	DESCRIPTION
TOP SOIL			TOPSOIL
RESIDUAL SOIL	VI		ALL ROCK MATERIAL IS CONVERTED TO SOIL
COMPLETELY WEATHERED	V		ALL ROCK MATERIAL IS DECOMPOSED AND/OR DISINTEGRATED TO SOIL
HIGHLY WEATHERED	IV		MORE THAN 35% OF THE ROCK MATERIAL IS DECOMPOSED TO SOIL
MODERATELY WEATHERED	III		LESS THAN 35% OF THE ROCK MATERIAL IS DECOMPOSED TO SOIL
SLIGHTLY WEATHERED	II		DISCOLORATION INDICATES WEATHERING OF ROCK MATERIAL
FRESH	I		NO VISIBLE SIGN OF ROCK MATERIAL WEATHERING

WEATHERING GRADES & PROFILE FOR THE ROCK MASS

Fig. 4.9 Weathering grades and profile for the rock mass.

A fracture spacing should preferably be measured directly on a rock face, and the following is recommended (Anon, 1977):

	Fracture spacing index I_f (metres)
Extremely High	> 2
Very High	0.6–2
High	0.2–0.6
Medium	0.06–0.2
Low	0.02–0.06
Very Low	0.006–0.02
Extremely Low	< 0.006

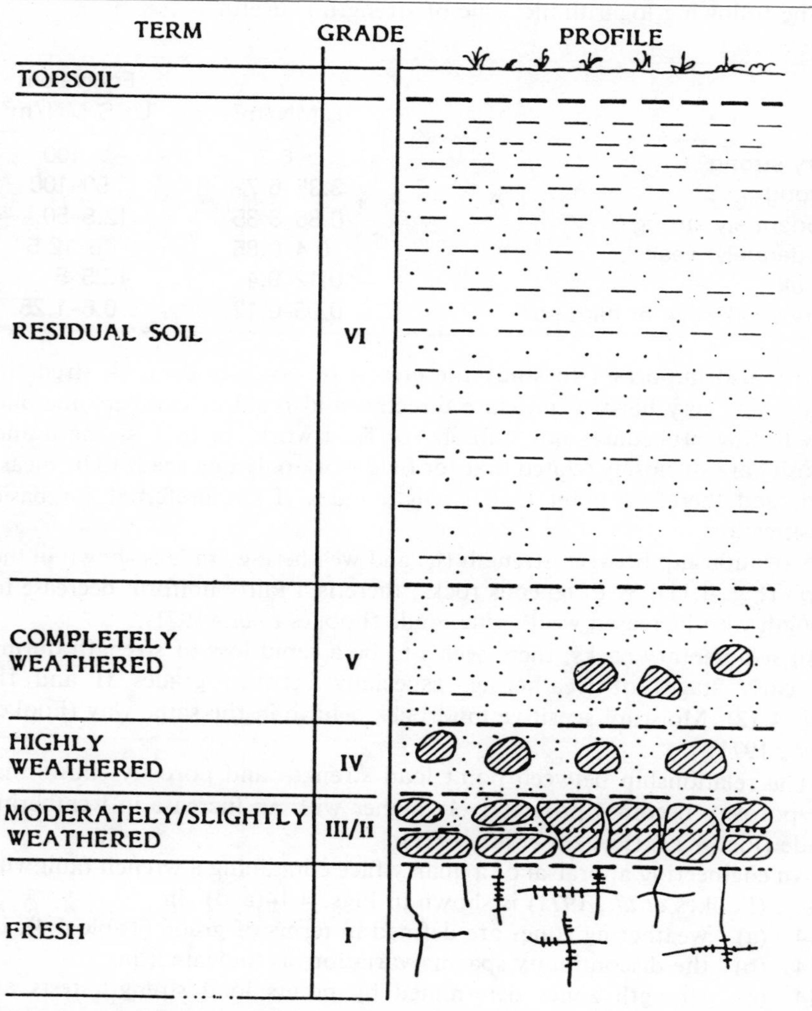

Fig. 4.10 Typical weathering profile over igneous rocks on low angled slopes and flat ground.

C. Strength classification

The point-load strength tester (Broch and Franklin, 1972) is a portable instrument suitable for field estimation of rock strength. Its advantages are:

(a) produces tensional failure, i.e. lower loads than for compressive failure
(b) platen contact conditions are of little significance
(c) irregular test lumps can be used
(d) large numbers of tests can be carried out quickly to determine a range of strength for a particular rock type.

The following logarithmic scale of strength is useful:

	I_s MN/m²	Equivalent UCS MN/m²
Very strong	> 6.7	> 100
Strong	3.35-6.7	50-100
Moderately strong	0.85-3.35	12.5-50
Moderately weak	0.4-0.85	5-12.5
Weak	0.12-0.4	1.25-5
Very weak rock or hard soil	0.05-0.12	0.6-1.25

It is also important to study the effects of porosity on rock strength. Porosity is very hard to measure directly, and requires cumbersome and slow testing procedures not suitable for field work. In fact, strength and porosity are so closely related that for field work only one needs to be measured, and therefore point load strength index (I_S) is preferred for basic classification.

A relationship between strength (I_S) and weathering grade is shown in the graph (Fig. 4.11). With igneous rocks, there is a fairly uniform decrease in strength with increased weathering grade (Fookes *et al.*, 1971).

In sedimentary rocks, there seems to be a rapid loss of strength during the early stages of weathering, especially between grades II and III (Fig. 4.12). Moisture sensitive mudrocks behave in the same way (Fookes *et al.*, 1971).

The relationship between point load strength and porosity shows that the porosity becomes progressively higher with an increase in weathering grade (Fig. 4.13) (Fookes *et al.*, 1971).

An engineering appraisal of a quarry face containing a wrench fault with gouge (Fookes *et al.*, 1971) is shown in Figs. 4.14(a–d). In

4.14 (a) weathering zones are defined in terms of grade (Table 4.4); in
4.14 (b) the discontinuity spacing variations is indicated; in
4.14 (c) strength zones determined by points load strength tests are shown; and in
4.14 (d) an engineering appraisal is given in terms of ease of excavation, in which each of the factors is taken into account.

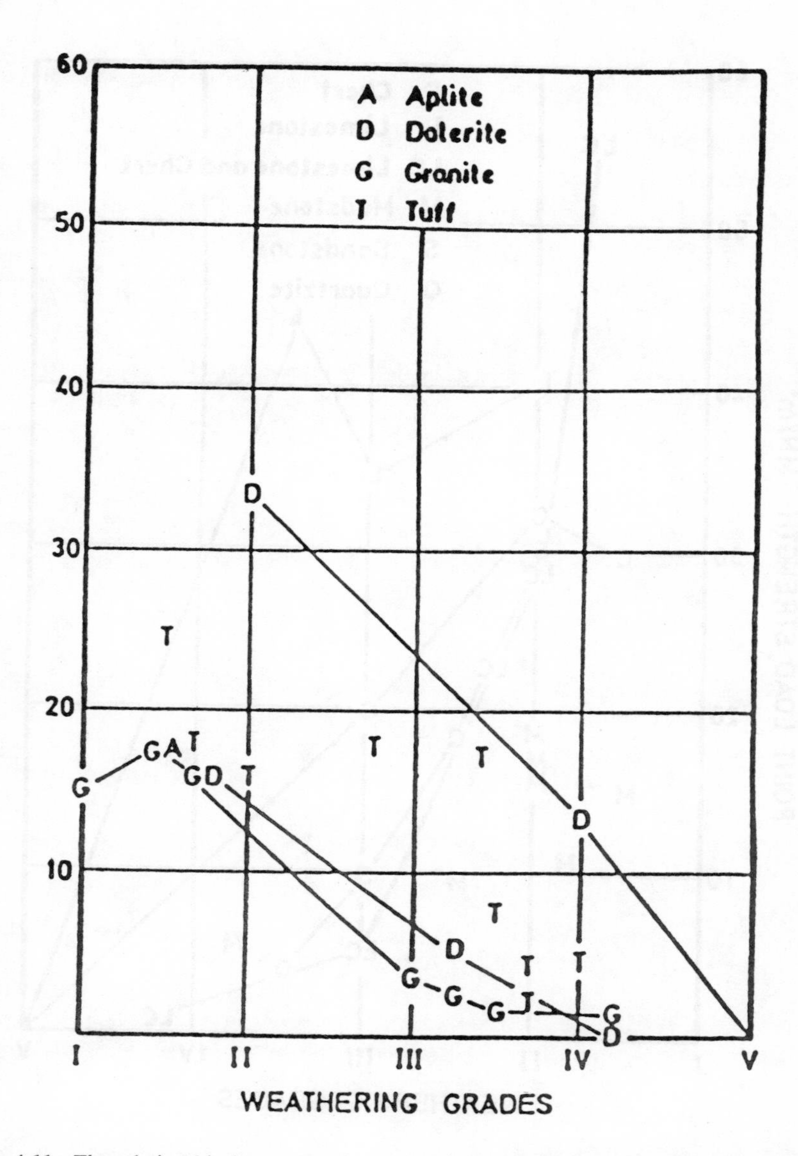

Fig. 4.11 The relationship between rock strength and weathering grade for certain igneous rocks. (Fookes, P.G., Dearman, W.R. and Franklin, J.A., 1971; Copyright © 1971 by *Geological Society of London*)

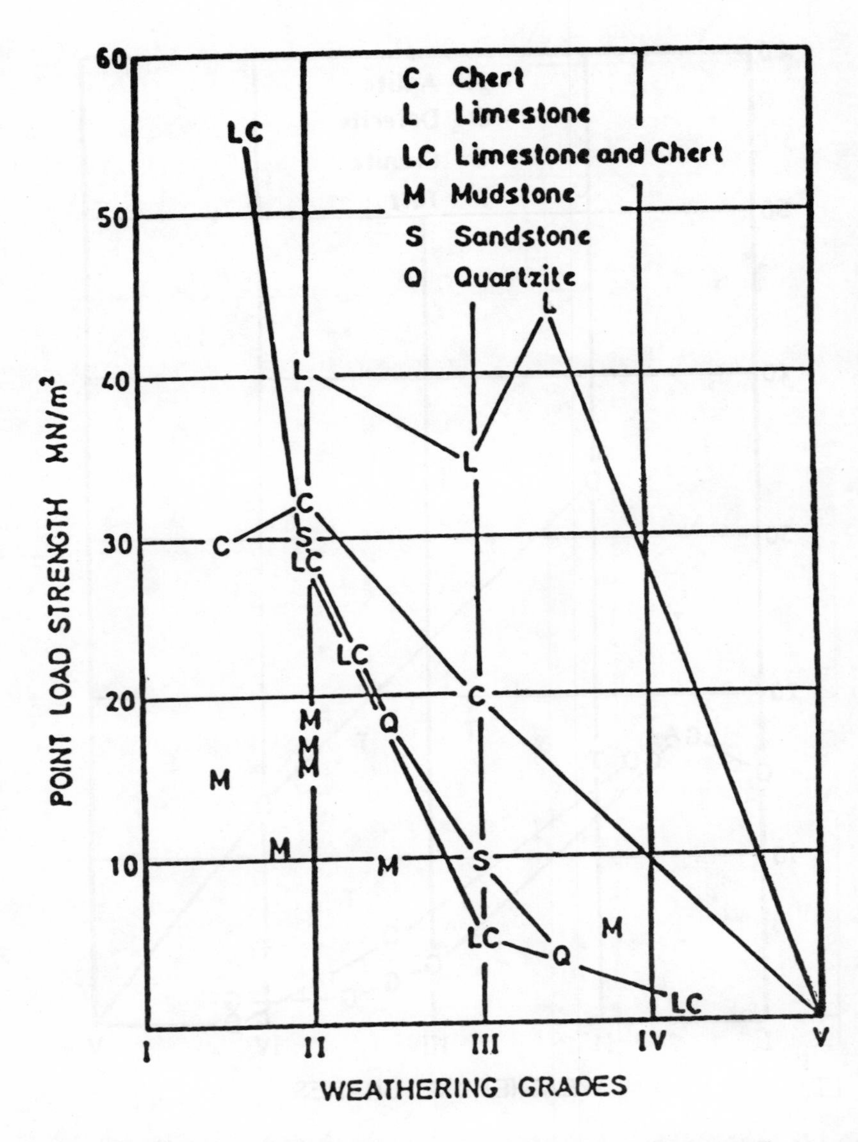

Fig. 4.12 The relationship between rock strength and weathering grade for certain sedimentary rocks. (Fookes, P.G., Dearman, W.R. and Franklin, J.A., 1971; Copyright © 1971 by *Geological Society of London*)

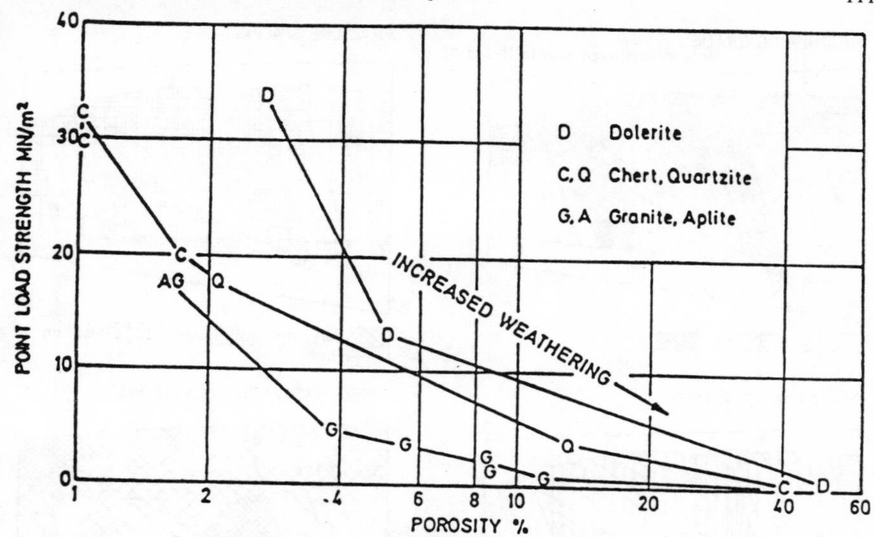

Fig. 4.13 The relationship between rock strength, porosity and intensity of weathering for a variety of rock types. (Fookes, P.G., Dearman, W.R. and Franklin, J.A., 1971; Copyright © 1971 by *Geological Society of London*)

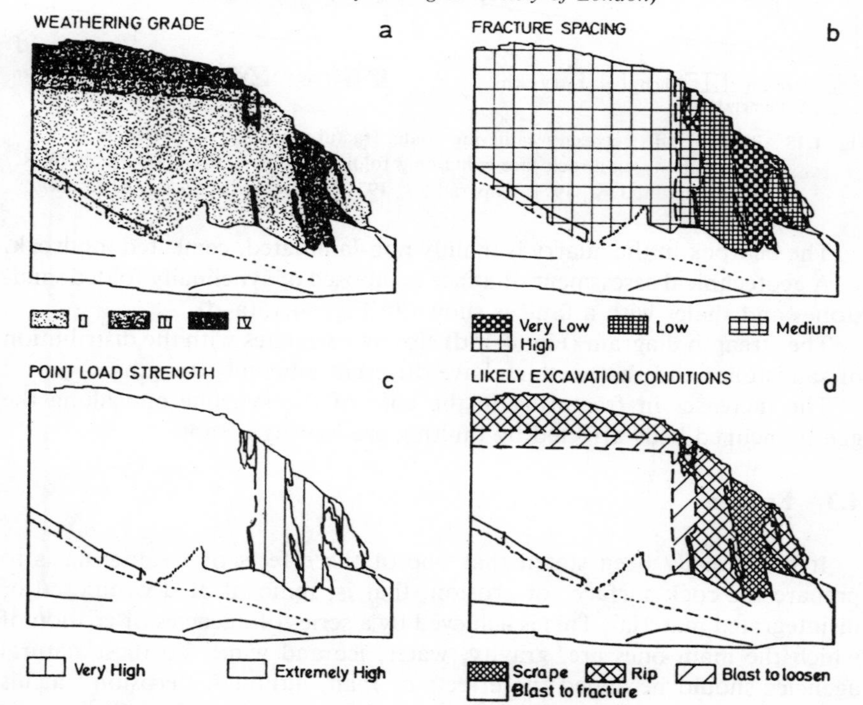

Fig. 4.14 Relationships between weathering grade, fracture spacing, rock strength and excavatability on a rock face containing a fault. (Fookes, P.G., Dearman, W.R. and Franklin, J.A., 1971; Copyright © 1971 by *Geological Society of London*)

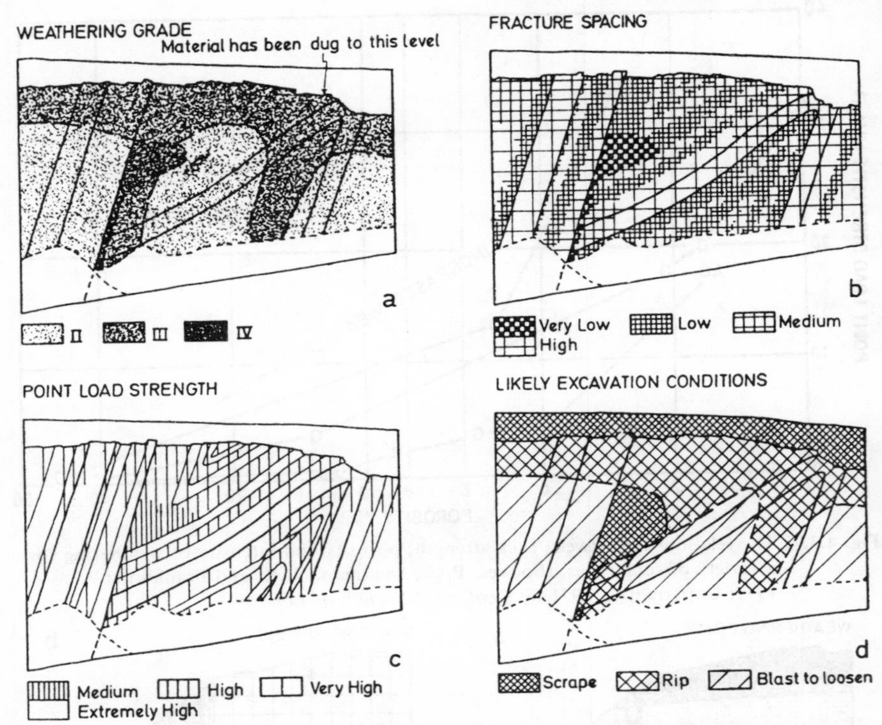

Fig. 4.15 Relationships between weathering grade, fracture spacing, rock strength and excavatibility on a rock face containing folding. (Fookes, P.G., Dearman, W.R. and Franklin, J.A., 1971; Copyright © 1971 by *Geological Society of London*)

The bedrock in the quarry is mainly non-laminated, cemented mudrock. A geotechnical assessment of a face composed of synclinally folded sandstones and shales with a fault is shown in Figs. 4.15(a–d).

The strength diagram (Fig. 4.15d) closely correlates with the distribution of sandstones and shales which have different inherent strengths.

The increases in fracturing in the core of the syncline and along the gently inclined beds affected by faulting are worthy of note.

4.3 Erosion

It has already been stated that one of the effects of weathering is to prepare the rock surface for erosion, that is, removal of decomposed or disintegrated material. This is achieved by a series of agencies of erosion of which the main ones are, gravity, water, ice and wind. To these natural agencies should be added the effects of man, although "erosion" at his hands does not always require weathering to have taken place first.

Erosion by gravitational agencies covers mainly the field of mass movement on slopes, that is, landslides and related phenomena. A variety of classifications exist dealing with mass movements, and Table 4.5 is an example of one based on mechanism of movement and types of material involved (Varnes, 1978).

Table 4.5 (from Varnes, 1978)

TYPE OF MOVEMENT			TYPE OF MATERIAL		
			BEDROCK	ENGINEERING SOILS	
				Predominantly coarse	Predominantly fine
FALLS			Rock fall	Debris fall	Earth fall
TOPPLES			Rock topple	Debris topple	Earth topple
SLIDES	ROTATIONAL		Rock slump	Debris slump	Earth slump
	TRANSLATIONAL	FEW UNITS	Rock block slide	Debris block slide	Earth block slide
		MANY UNITS	Rock slide	Debris slide	Earth slide
LATERAL SPREADS			Rock spread	Debris spread	Earth spread
FLOWS			Rock flow (deep creep)	Debris flow (soil creep)	Earth flow (soil creep)
COMPLEX			Combination of two or more principal types of movements		

Falls involve the free fall of blocks of soil or rock of any size without significant amounts of shear displacement (Fig. 4.16a). The surfaces which delineate the extent of the dislodged mass are pre-existing discontinuities.

Toppling failure has only been relatively recently recognized as an important mechanism of landsliding. It involves the forward rotation of rock masses with a generally columnar structure or with dominant discontinuities dipping into the slopes at a high angle. Frequently, shear failure at the toe of the slope removes sufficient lateral confinement for the backing columns to rotate forwards (Fig. 4.16b).

True landslides involve shear strain and displacements over single or multiple shear surfaces. The group is very complex and covers a variety of individual failure modes. The most important of these are rotational failures (Fig. 4.16c) where the failure surface is distinctly cylindrical or spoon-shaped, and planar, where failure is along a pre-existing plane surface (Fig. 4.16d). The other main subdivision is into slides in which a large amount of deformation of the sliding mass takes place and ones where the slipped block remains substantially intact. (Figs. 4.16d, i and ii).

Lateral spreads are not very common types of failures, and generally result in a marked extension of a failing mass without the development of

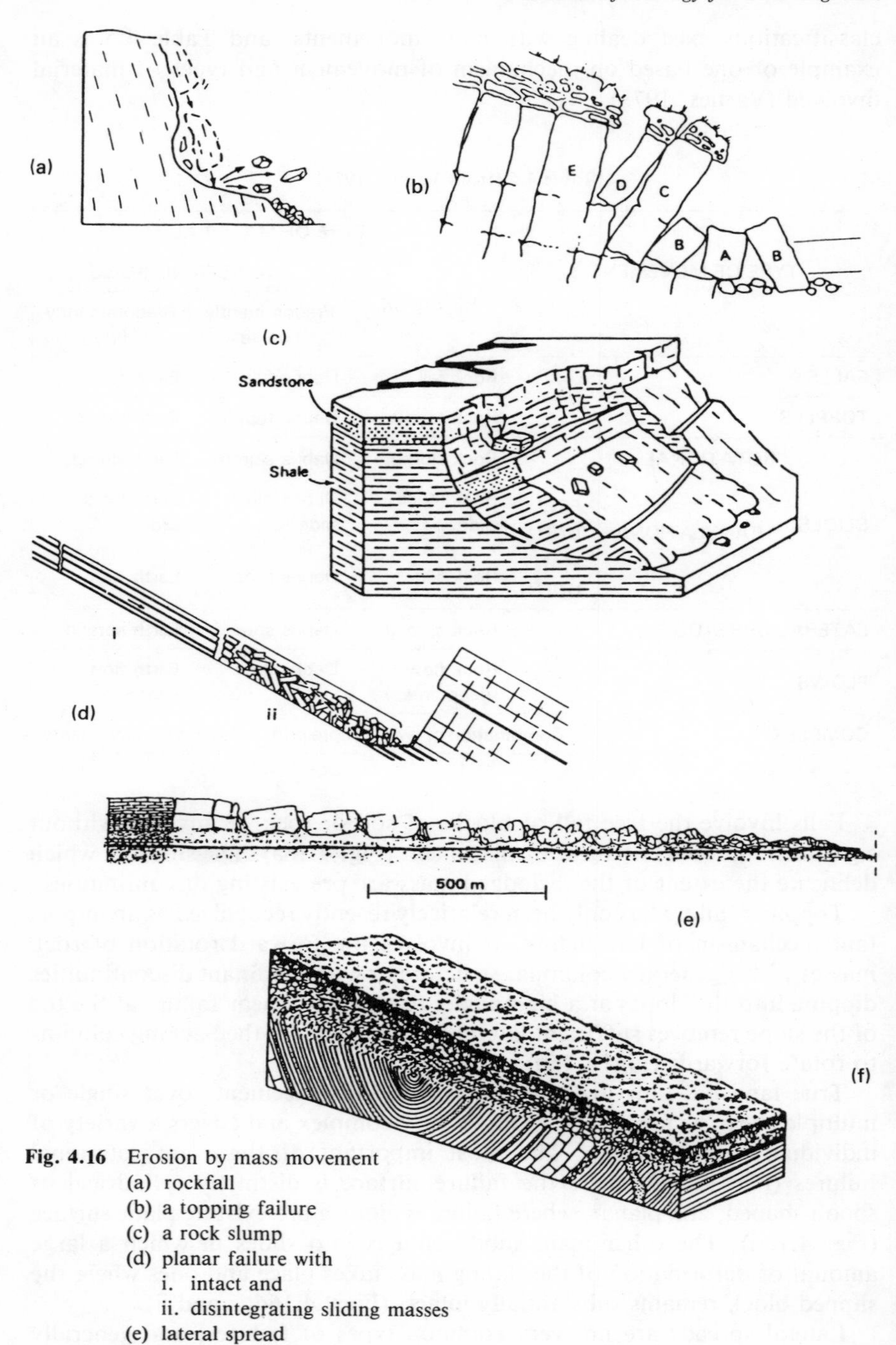

Fig. 4.16 Erosion by mass movement
(a) rockfall
(b) a topping failure
(c) a rock slump
(d) planar failure with
  i. intact and
  ii. disintegrating sliding masses
(e) lateral spread
(f) slope creep

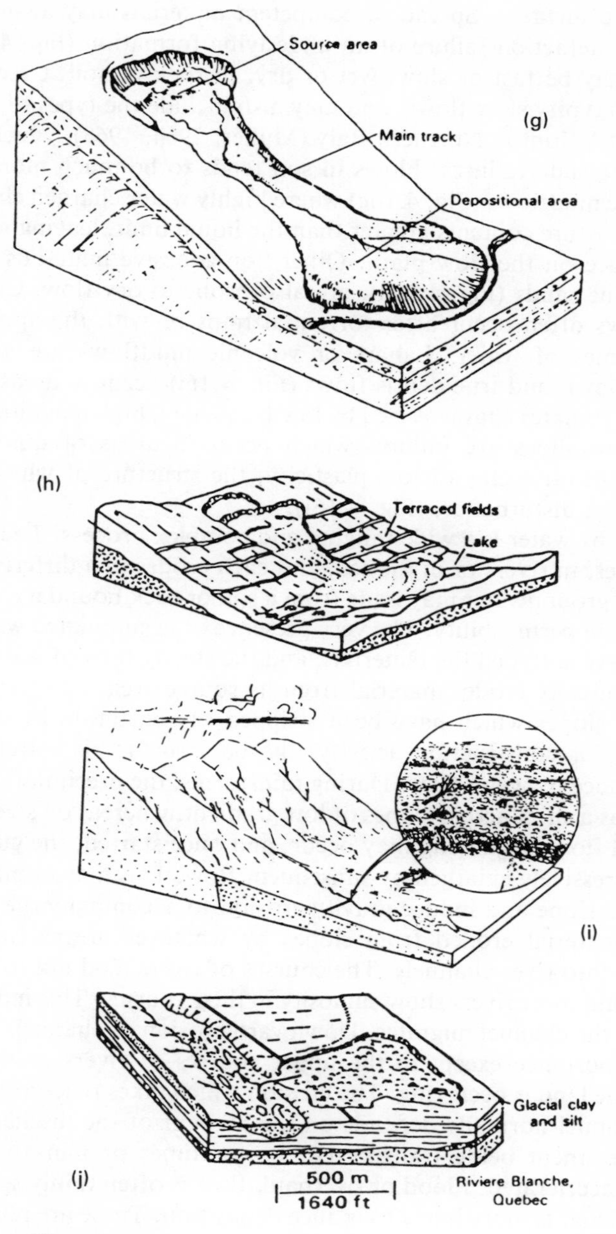

(g) mudflows
(h) sand flows
(i) debris flows
(j) flow slides
(Varnes, D.J., 1978; Copyright © 1978 by the Transportation Research Board, National Academy of Sciences, Washington, D.C.)

clear failure surfaces. Spread of competent materials may also occur as a result of liquefaction failure of an underlying formation (Fig. 4.16e).

Flows may be fast or slow, wet or dry, in rock or soil. Creep motions (Fig. 4.16f) typify slow flows, and may also include the types of movement observed at Vaiont in Northern Italy (Muller, 1964, 1968), which preceded catastrophic slide failures. Flows in soil tends to be much more common and include mudflows (Fig. 4.16g) where highly water charged clayey debris flows at moisture contents greater than the liquid limit, leaving no sign of a shear surface on the flow plath. Other non-cohesive materials, especially silts and fine sands (Fig. 4.16h), are also prone to dry flow. Catastrophic debris flows often occur after torrential rains or with the application of large volumes of water. Lahars or volcanic mudflows are a particular danger in Java, and true debris flows (Fig. 4.16i) occur in areas with steep slopes and climates characterized by low frequency, high magnitude precipitation. Flow slides are failures which occur in areas of quick clay, an extremely sensitive clay of low plasticity, the structure of which collapses (liquefies) on disturbance, (Fig. 4.16j).

Erosion by water is a wide ranging and complex process. The water may be on slopes, in rivers or in the sea, and each erodes in a different fashion. On slopes, groundwater may issue from a soil or rock boundary where there is a change in permeability. The very presence of accumulated water there is likely to have softened the materials, and the steady flow of water from the ground gradually erodes material from its source area.

On soil slopes which have been stripped of vegetation by mass movement, fire, or construction activity, the bare surface is extremely prone to disturbance by water flow. During rainstorms, the precipitation will tend to runoff as a sheet across slopes of low gradient, whereas on steeper slopes, channelled flow or gullying may occur, and once started, the gully is likely to be progressively enlarged by subsequent flow. This may eventually oversteepen the slope and induce gravity erosion to accompany the gullying.

Much material eroded from slopes by whatever means finds its way eventually into river channels. The courses of rivers tend not to be static in position, and most rivers show sinuosity in their courses. This induces lateral erosion as the channel migrates. Downward erosion of channels is generally of less importance except in the upper courses of rivers or where active uplift of the land is occurring. The erosion which takes place at any point in a river channel normally depends on the amount of the discharge and the load of sediment being transported. So, at times of bankful discharge, erosion is accelerated. Flood or overbank flow is often thinly spread over a floodplain and is more likely to induce deposition. These are relatively high magnitude events of low frequency, and should be compared with "normal" events to assess their relative importance. In many humid climates, normal flows (high frequency, low magnitude) achieve more in terms of erosion than do abnormal flows.

Marine erosion (Clark, 1979) depends largely on wave action, and the more erosive breakers are usually generated by onshore winds blowing over large expanses of sea (fetch). The dominant wave type will largely determine whether a beach is gently or steeply inclined (Glavin, 1968). High amplitude and high frequency (13 to 15 per minute) waves are generally associated with steep beaches. They have nearly circular water motion but end in quite distinct ways. The first type, on breaking, produces a crest which plunges vertically downwards affecting a relatively narrow zone of the beach (Fig. 4.17a). The swash, or rush up the beach is weak, and interference from backwash from the previous wave is high because of the high frequency. The second type, a surging breaker, does not seem to really break at all, and rushes up the slope of the beach, (Fig. 4.17b). In contrast, some waves are low amplitude, low frequency (6 to 8 per minute) waves which have elliptical water movements. On breaking, they spill forward, moving material up the beach in the swash, whilst the backwash is weak since little interference from preceding waves occurs with the low frequency of arrival (Fig. 4.17c).

Breaking of waves probably occurs both as a result of over steepening of the waves, generation of turbulence within a wave, and friction with the sea bed. Breaking tends to occur when the ratio between the depth of water and the height of wave is between 1.1 and 1.5. After breaking, the wave flows turbulently up the shore, although its erosive influence may be checked both by interference from the backwash of preceding waves and by infiltration.

The wind creates wave motion and changing wind patterns can alter the pattern of wave attack as explained in Chapter 1. Violent storms can cause waves to act at higher levels, with sea levels increased by two to three metres. These storm surges are particularly important when dealing with low lying coastal areas, and if the effect coincides with a period of high tide, it can cause breaching of sea defences and catastrophic flooding.

Wind only really becomes an important agent of erosion in arid and semi-arid areas, although in some humid areas, the effects of wind can be accelerated by changing the thresholds of potentially susceptible materials. Over-drainage of land, excessive ploughing and break-up of material are examples of changing materials, and removal of hedgerows and trees of changing the unbroken distance over which the wind blows. The main processes of erosion by wind are deflation and abrasion. Deflation is the removal of fine sand and smaller particles by the wind. It may occur in unvegetated regions or areas of poor agricultural practice. Hollows or "blow-outs" may result from this. Erosion of part of a surface deposit may occur, where fines are blown away leaving the area covered by a gravelly deposit. Abrasion is effected by wind-driven sand grains and is particularly influential at the base of steep slopes.

Although glaciers are less important than rivers in terms of overall erosion on a world scale, they have shaped much of the landscape of Europe and North America. Pure ice on its own is a very ineffective tool for eroding massive rocks. Rock fragments become incorporated into the moving ice mass; and it is thus embedded debris and the physical motion of the ice

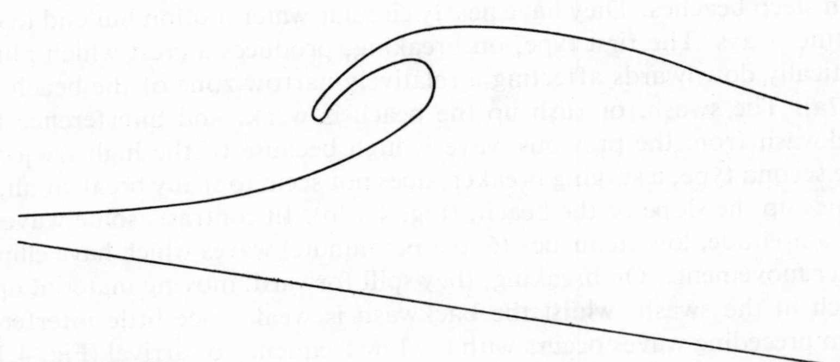

(a) Plunging breaker.

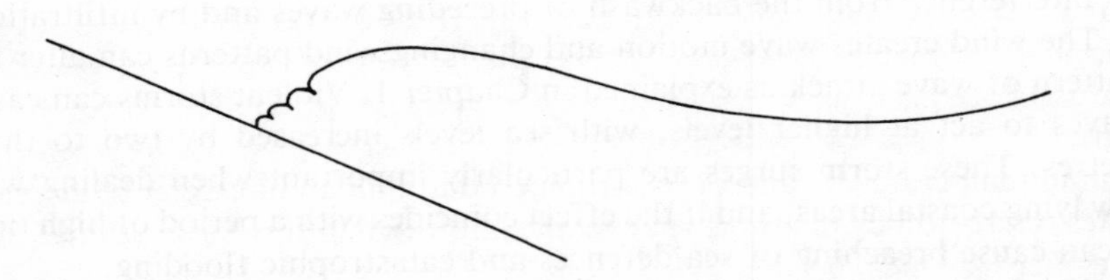

(b) Surging breaker, steep bottom slope.

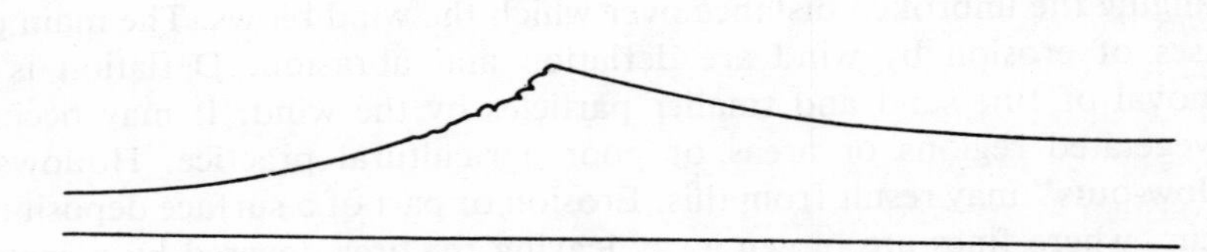

(c) Spilling breaker, gentle bottom slope.

Fig. 4.17 Main types of breaker,
(a) plunging
(b) surging and
(c) spilling
(Glavin, C.J., 1968; Copyright © 1968 by *American Geophysical Union*)

which makes glaciers so potent an erosive force. The ice may work its way into existing fractures in a rock and drag a block of rock away from the mass, or it may scrape and scratch on rock surfaces usually smoothing and polishing the rock. Furthermore, actual crushing and fracturing of bedrock by ice renders it more susceptible to later erosion. Ice occurs as major sheets or as valley glaciers, and each has shown itself to be a potent erosive force.

The erosive effects of man are numerous and are of both a direct and indirect nature. Excavation of all kinds whether for construction works or mineral extraction is an obvious example of a direct kind. Indirectly, the results of industrialization and urbanization and the emission of millions of tons of carbon dioxide and other compounds into the atmosphere has made rainfall much more acidic. This has greatly accelerated the rate of decomposition of rocks, either in their natural state or in the form of building stones. The effects of ploughing and breaking up the soil, increasing the area of tiles and concrete or tarmac and hence the amount of runoff, and taking out most of the sediment of rivers at reservoir sites have already been mentioned as having a significant effect on amounts and patterns of erosion. These accelerated rates have been going on for 8000 years at least and as techniques become larger scale and more efficient, the effects are more acute. The use of nuclear explosions for engineering excavation has already commenced, and if increased will represent a further acceleration in the rate of change. The effects tend to be greater where agriculture and industry are practised in an intensive way. In semi-arid lands, the effects of agricultural practice are particularly severe in increasing rates of erosion. In humid temperate latitudes, it is estimated that human interference increases rates of erosion by ten times on average, although in some exceptional cases the increase may be a hundred-fold. In humid tropical areas, although details are less well known, the same pattern emerges. In Java for example, sediment carried in rivers in 1911 revealed an average erosion of 900 m^3 per km^2. After a programme of deforestation in 1934 this increased to 1900 m^3 per km^2.

Once weathered debris has been picked up by an agent of erosion, that agent in effect becomes an agent of transportation. It redistributes weathered material to a new locality either close by, as is usually the case with a slope failure, or to some distant location, as a river or glacier might do. The modes of transportation are complex, and it is not necessary to go into them here in any detail. For the most part though, of particular importance is the sorting of material which goes on during the transportation process. Sorting of weathered debris is in terms of particle size, and the spread of particle sizes present at any point in, say, the cross section of a river, is a reflection of the carrying capacity at that point. In general, carrying capacity is greater if flow is turbulent and able to keep a fragment suspended in the fluid medium, but suspension is not a prerequisite for movement. The main types of movement of material by water and wind are similar and may be summarized as:

(1) drag or rolling, mainly of the coarsest fraction present;
(2) saltation or bouncing of the next finest particles;
(3) suspension of the finest particles within the fluid medium.

In addition, solution may be an important mode of transportation in water. Generally, ice sorts debris only crudely if at all, carrying it at the base, within and on top of the ice. It is usually the case that true glacial deposits as opposed to those resulting from meltwater, are badly sorted and well graded, with particle size distributions ranging from boulders to clay.

Deposition of material takes place when there is a decrease in the carrying capacity of the transporting medium. The amount by which it is decreased is directly reflected in the size of sediment deposited. In marine conditions in the continental shelf, sediment generally becomes gradually finer offshore. It reflects the available energy for transporting coarse size material by breaking waves near and on the beach, and the rapid decrease in that energy as the water becomes deeper. The resulting sediment zones tend to be fairly uniform and continuous. On land, this is certainly not the case, and deposition from rivers, glaciers, by the wind or on slopes occurs in a much more unpredictable and changeable way. In each case, sediments change rapidly both laterally and vertically and are associated with individual features of landscape and the changes in the regime of the transporting medium. It therefore helps to know something about the processes which act in each of the main transporting media so that the nature of possible changes may be predicted.

In rivers, so much depends on the state of flow in the river at a particular time. As flow varies then so does the sediment deposited. This pattern extends laterally as well as vertically and is complicated by seasonality in the climatic regime and therefore in flow. The type of pattern which emerges is complex, as indicated for part of the Nam Mune in Thailand (Fig. 4.18) (Leopold *et al.*, 1964). When this ceases to exist as a river system, the pattern of sedimentation will remain as a complicated and changeable sequence, and the deposits of the Old Alluvium of Singapore laid down up to two or three million years ago may be a deposit roughly equivalent to those of the Nam Mune today (Fig. 4.19) (Gupta *et al.*, 1980).

Seasonality and its influence on regime is perhaps even greater with glaciers. The range of deposition is vast and includes three major modes from the ice itself, deposition by meltwater streams and deposition from ice or meltwater streams into meltwater lakes or the sea. Non-uniform and irregular conditions are the norm, and the range of sediments is very great. They include tills which may range from normally consolidated to greatly over consolidated bouldery, gravelly, sandy, silty clays, to well sorted clays, sands, silts and gravels. Locally repeated advances and retreats of the ice further complicate and disturb this pattern. Glacial deposits are a common feature of North America, Northern Europe and Northern Asia and resulted from the major glaciations which ended approximately 10,000 years ago.

The various agents of weathering, erosion, transportation and deposition

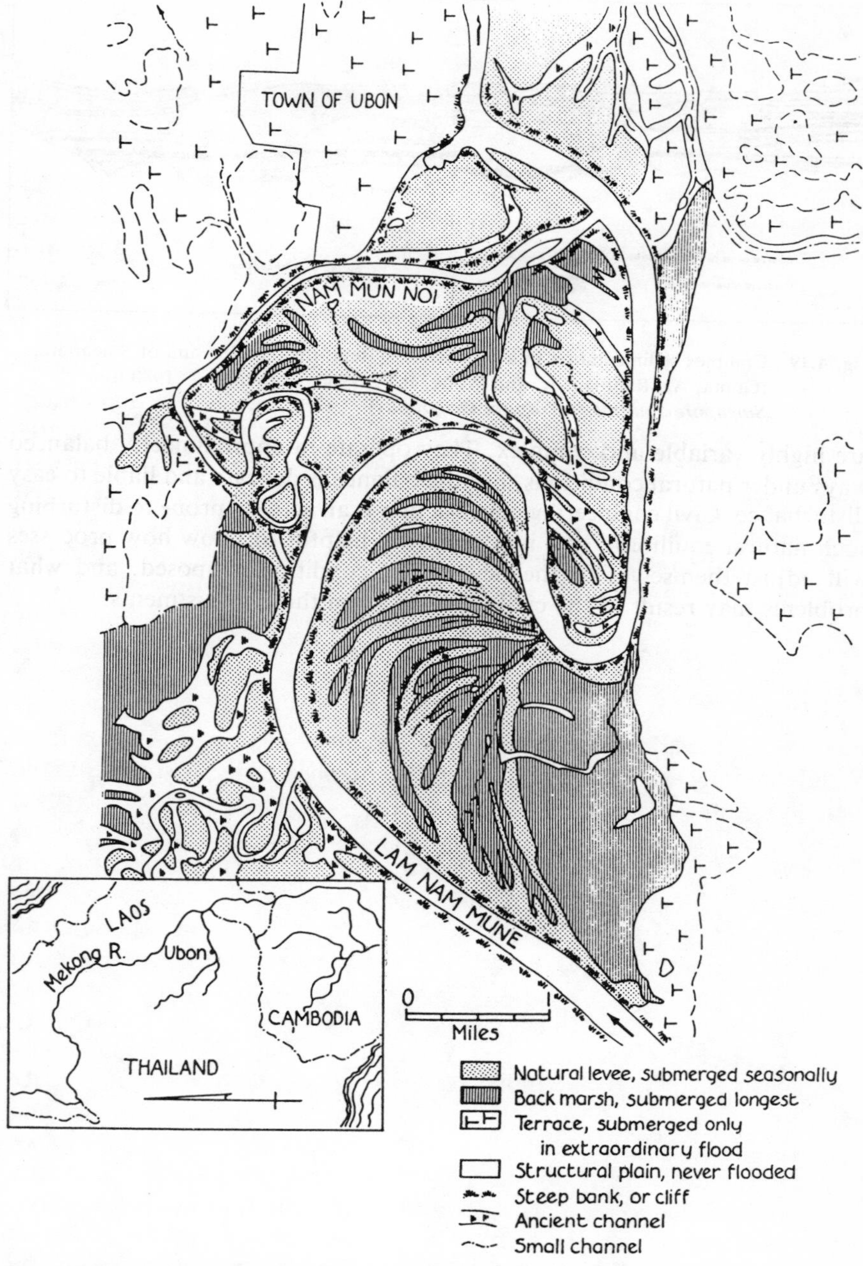

Fig. 4.18 Complex sedimentation pattern in a contemporary river system: Lam Nam Mune, Thailand. (From: Fluvial Process in Geomorphology by Leopold, L.B., Wolman, M.G. and Miller, J.P.; Copyright © 1964 by W.H. Freeman and Company. All rights reserved.)

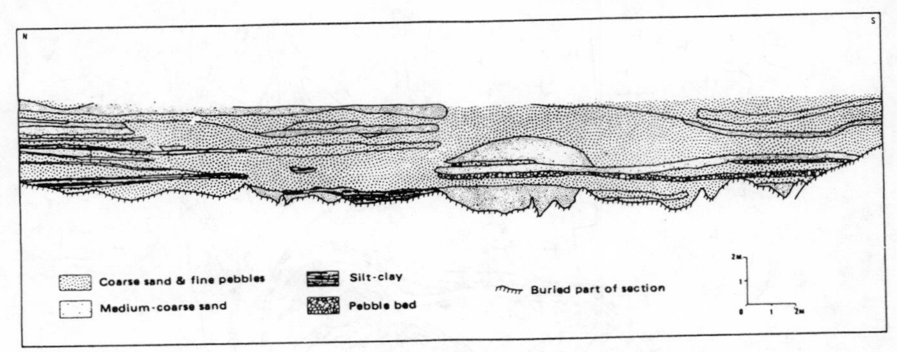

Fig. 4.19 Complex sedimentation in a fossil river system: the old alluvium of Singapore. (Gupta, A., Rahman, A. and Wong, P.P., 1980; Copyright © 1980 by *Singapore Journal of Tropical Geography*)

are highly variable and complex. They operate in approximately balanced ways under natural conditions, but the balance is delicate and liable to easy disturbance. Civil engineering construction is an activity prone to disturbing such natural equilibria, and it is therefore helpful to know how processes will adjust themselves to the new set of conditions imposed, and what problems may result to the civil engineer from those adjustments.

CHAPTER 5: SOIL PARTICLES, SOIL FABRICS AND SOIL STRUCTURES

Soil is a non-indurated aggregation of mineral and/or organic particles with air and/or water filled voids. Most soils result from the agencies of weathering, erosion, transportation and deposition. Organic soils such as peat, are not of this type, usually being an assemblage of vegetable remains which have not been destroyed by oxidation.

Soils are primarily classified on the basis of particle size. Each of the particles considered will therefore fall into a prescribed size range, and will form a soil which is represented by the dominant particle size. Particle size is an easy parameter to measure and controls many aspects of the engineering behaviour of a soil. The particle size ranges adopted in civil engineering are usually as follows:

1. Boulders, > 200 mm
2. Cobbles, 60 to 200 mm
3. Gravel, 2 to 60 mm, subdivided into
   - (i) coarse gravel, 20 to 60 mm
   - (ii) medium gravel, 6 to 20 mm
   - (iii) fine gravel, 2 to 6 mm
4. Sand, 0.06 to 2 mm, subdivided into
   - (i) coarse sand, 0.6 to 2 mm
   - (ii) medium sand, 0.2 to 0.6 mm
   - (iii) fine sand, 0.06 to 0.2 mm
5. Silt, 0.002 to 0.06 mm, subdivided into
   - (i) coarse silt, 0.02 to 0.06 mm
   - (ii) medium silt, 0.006 to 0.02 mm
   - (iii) fine silt, 0.002 to 0.006 mm
6. Clay, < 0.002 mm

The sands and gravel are cohesionless particles, that is they possess no inter-particle bond. Clays and to a much lesser extent silts are usually cohesive, but it should be remembered that "clay" describes only the dominant particle size found in the deposit. It does not automatically mean that the clays are cohesive or plastic, that is, composed of clay minerals. Some clays are formed from the grinding down of common rock forming minerals by glacial action and have a low plasticity and a very unstable, collapsable structure.

Plasticity is also used as a basis of classification of soils. The fine fraction, of both fine-grained and dominantly coarse-grained rocks, is tested for plasticity characteristics. The tests are the Atterberg Limit tests, measures of the liquid and plastic limits and hence, the plasticity index of the soil. The plastic limit is the water content of a soil as it just begins to behave in a plastic manner. The liquid limit is the water content when a soil just begins to behave as a liquid. The plasticity index is the range of moisture content over which the soil behaves in a plastic manner and is expressed as,

Plasticity Index (P.I.) = Liquid Limit (W_L) − Plastic Limit (W_P)

5.1 Non-cohesive Particles

Particles of gravel size are usually but not always composed of resistant minerals like quartz, flint, chert, or very resistant rock fragments, say of quartzite. This is not always the case however, as weathering subsequent to deposition, and the relief of an area may have a significant influence on the proportion of unstable minerals present in the deposit. In the case of low relief, only inert residual particles tend to remain and little material of such a coarse grain tends to be produced. In areas of high relief, the mineral assemblage in a gravel deposit may closely reflect the source of the rock from which it was derived. The rapidity of the processes in such areas may cause a variety of unstable grains to be preserved, as they existed in the parent rock. Subsequent weathering, erosion and redistribution to areas of progressively lesser relief will increase the maturity of the deposit.

The properties of coarse-grained soils depend on a number of geological factors which include particle shape (Fig. 5.1) and packing (Fig. 5.2). The former is mainly dependent on the length of the transportation history, and the latter on the rate of deposition and subsequent processes of compaction. Tight packing of grains leaves relatively small pore spaces, and the openness of the soil is low if the range of particle sizes is great. In soils composed of equidimensional grains however, a wide range of openness of fabric exists, and this is described by the soil's void ratio (e), the ratio of the volume of voids to the volume of solids. A range of e values of between 0.35 and 1.0

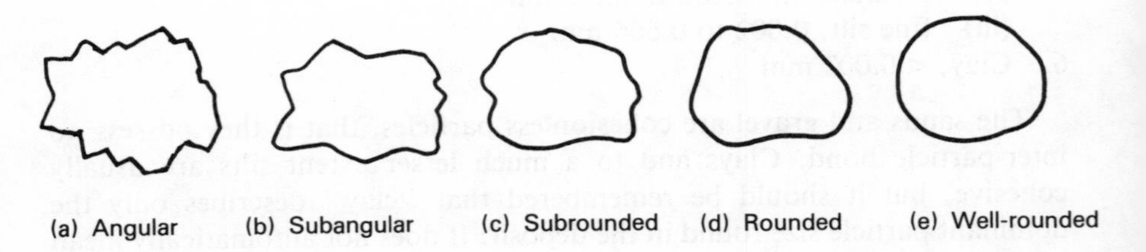

Fig. 5.1 Degree of rounding of soil particles.

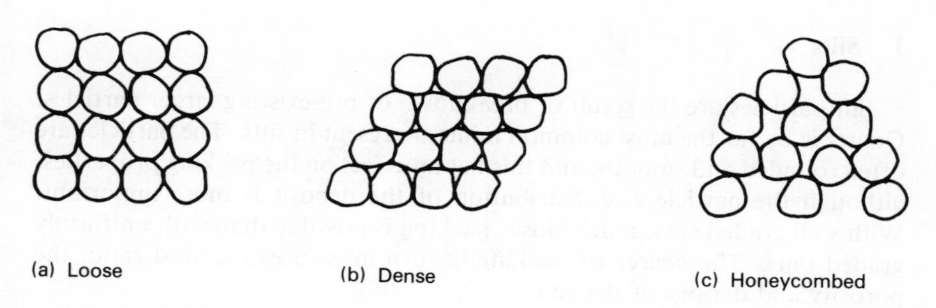

Fig. 5.2 Packing of soil particles.

may exist and the soil skeleton remain stable. If an e value of > 1.0 is recorded, then the fabric may be collapsable, the soil possessing an extremely open, honeycombed fabric with a "delicately balanced" packing of grains.

Shape and packing have an influence on the strength of a granular deposit. The strength of a soil largely depends on the number and area of points of contact between grains in the soil. The frictional strength of the soil varies as shown in Table 5.1. With angular grains, more interlock occurs between the grains during shear than with rounded grains, adding to the frictional resistance. However, this effect is less clear with gravels as crushing is more important at the points of contact. In any type of granular deposit, the crushing which occurs at any mean particle size will be greater in uniformly sorted soil than in a well graded one since the larger number of inter-particle contacts in the latter will decrease the load per contact.

Table 5.1 Effect of grain shape and grading on the peak friction angle of cohesionless soils.

SHAPE AND GRADING	LOOSE	DENSE
1. Rounded, uniform	30°	37°
2. Rounded, well graded	34°	40°
3. Angular, uniform	35°	43°
4. Angular, well graded	39°	45°

5.2 Cohesive Particles

In this group, silt and clay particles are to be included. Some qualification is necessary before describing the particles themselves, and that is, that the individual particles which are derived from pre-existing rocks by breakdown and wear are not in themselves cohesive. It is only in the mass that a cohesion and a certain amount of plasticity become apparent. However, with clay minerals, some of which may be larger than 0.002 mm in size, true cohesion and plasticity are true characteristics of the particles.

1 Silts

Silt particles are the result of breakdown of pre-existing larger particles. Quartz is by far the most common mineral present in silts. The particles are often rounded and smooth, and this has an effect on the packing properties, although the particle size distribution of the deposit is more important. With well graded silts, much closer packing is possible than with uniformly graded ones. The degree of packing in turn influences the void ratio, the porosity and density of the soil.

Silts laid down on flood plains, a common source of this size of material, tend to be well graded and stable in structure. Wind blown silts or loess have a very uniform grading which renders them far less stable, particularly from the point of view of settlement. Loess tends to have a very open structure which is prone to collapse on wetting, a process referred to as hydro-consolidation. The silt size fraction in loess varies between about 50% and 90%. At the lower end of the scale, much of the remainder of the soil is composed of clayey material, often with very few grain to grain contacts of the coarser material. Silts typically have a low liquid limit and plasticity index, averaging about $W_L = 30$ and IP = 4 to 9, although this very much depends on the proportion of clay material present in the matrix.

2 Clays

Although some common rock forming minerals, notably quartz, do occur in the clay size fraction, only the clay minerals will be dealt with here. Clay minerals are the most common products of decomposition of the

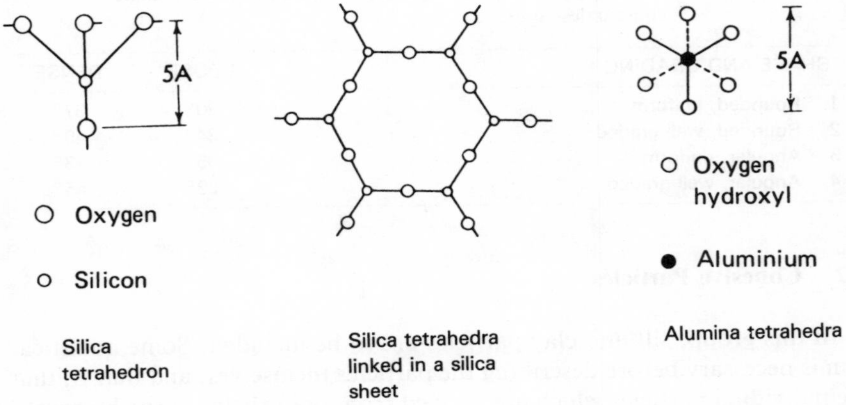

5A

Oxygen

Silicon

Silica tetrahedron

Silica tetrahedra linked in a silica sheet

5A

Oxygen hydroxyl

Aluminium

Alumina tetrahedra

Fig. 5.3 Basic molecular units of clay minerals.

common rock forming silicate minerals. Most of them have a layer lattice structure generally similar to that of micas, and their atomic structure consists of two basic units:

(i) silica tetrahedra, where four oxygen atoms surround a central silicon atom (Fig. 5.3a),

(ii) alumina octahedra in which units of Al are surrounded by oxygens and hydroxyls (OH), (Fig. 5.3b).

Most clay minerals consist of sheets of silica and alumina packed together to form the characteristic platy structure.

The main groups of clay minerals are:

1. Kaolinite
2. Illite
3. Smectites (Montmorillonites)

Kaolinites are formed from alternating layers of alumina and silica (Fig. 5.4a). The different members of the group arise from irregularities in the

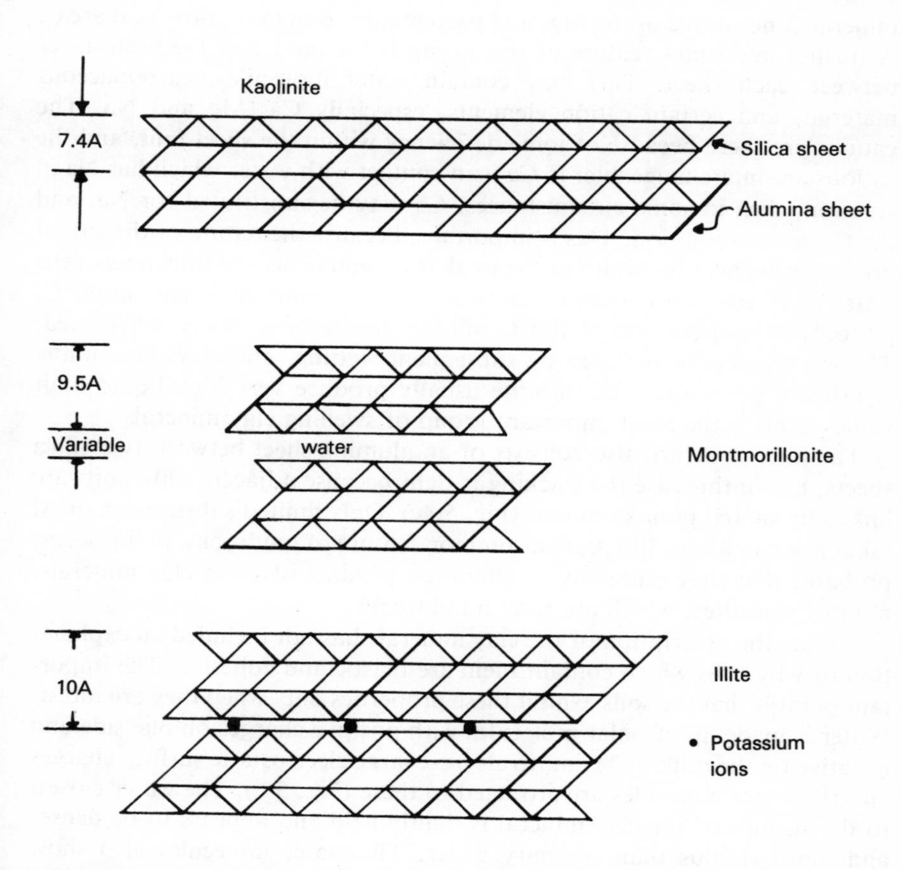

Fig. 5.4 Patterns of stacking of molecular units in different kinds of clay minerals.

alumina sheet which is commonly not in chemical balance, and which acquires different cations and establishes different linkages with the silica tetrahedra in order to achieve balance. The structure is generally stable and well bonded. One peculiar variation is when a layer of water is sandwiched between each pair of clay sheets. This increases the unit thickness to 10 Å from 7.4 Å . The water can be driven out by drying and the mineral may not revert to its hydrated form with wetting. This has been known to cause problems with such materials used in embankments. Properties measured on oven dry samples have not correlated with the natural, moist material used during construction. This mineral is called halloysite.

Smectites consist of one alumina sheet between two silica sheets (Fig. 5.4b). The unit thickness is 10 Å but this is extremely variable. Stacking is poor and easily disturbed. Extensive substitution of Mg for Al in the alumina sheets is a characteristic of this mineral group, and in theory, each substitution produces a new mineral. Smectites are commonly found as products of decomposition of rocks rich in ferromagnesian (dark) minerals, the source of the Mg, and particularly so in humid-tropical areas. A further important feature of this group is the nature of the weak layer between each sheet. This may contain water molecules, carbonaceous material, and certain cation elements, especially Ca, Mg and Na. The cations are there because of ionic deficiency within the sheet unit, and the cations are interchangeable. If Ca is in contact with water which has Na in solution, then in some circumstances, Ca may be substituted for Na, and the Ca go into solution. This is important, because the nature of the cation present influences the width of the weak layer and hence the thickness of the water layer which it contains. The water can be removed in part at 60 °C, but only at temperatures of 300 °C will the clay become totally dehydrated. The water is easily replaced by subsequent wetting, and clays containing significant percentages of smectite usually produce very high liquid limit values. This is the most important group of swelling clay minerals.

The illite structure also consists of an alumina sheet between two silica sheets, but in this case the stacking is tight because adjacent illite units are linked by shared potassium ions (Fig. 5.4c). Only limited substitution of Al takes place in illites. Illites are a common feature of mudrocks, and it seems probable that they represent an alteration product of other clay minerals, notably smectites, which are rare in mudrocks.

So far, the description of the clay minerals has not included an explanation of why soils which contain them are plastic and cohesive. The important point is that the soils exhibit these properties only when they are moist. Water is made up of polar molecules with positive charges on one side and negative on the other. Clay minerals have large electrostatic surface charges and the water molecules are attracted to these (Fig. 5.5). The water closest to the surface of the clay mineral is tightly held and appears to be denser and more viscous than ordinary water. The water molecules also show marked orientation with respect to the electrostatic field. Further away

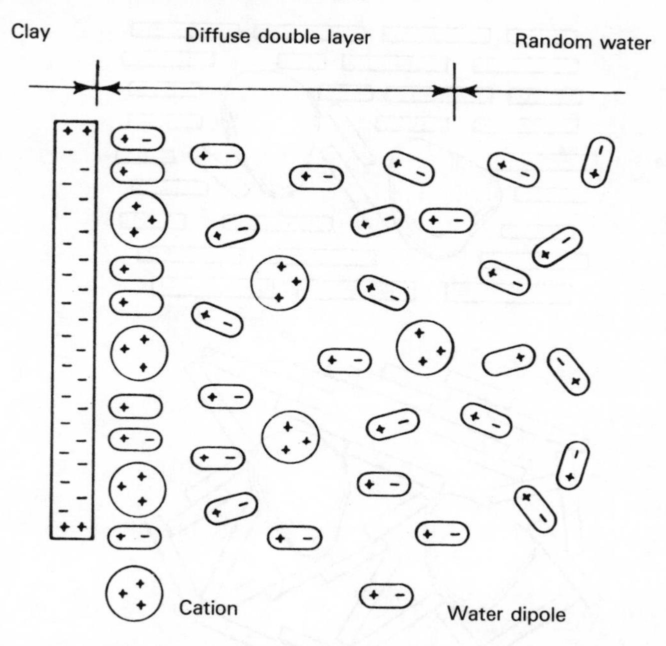

Fig. 5.5 Diagrammatic representation of the adsorbed water layer on the surface of clay minerals.

from the clay mineral surface, the bonding, viscosity and orientation becomes less and grade with increasing distance into random pore water. The tightly bonded layer is about 10 Å thick. The binding of water to the mineral surface is called adsorption, and the bound water, adsorbed water. It forms a link between one mineral particle and another, and binds them together. This permits adjustments of their relative positions when stressed, rather than a parting of the grains. As a result, clay, unlike sand can be moulded.

5.3 Clay Fabrics

The internal arrangement of particles within cohesive soils depends very much on the types of clay minerals presents and the forces between them. Some of the forces have been dealt with above, but of considerable significance is the nature of the water in which the clays were sedimented. This will influence the forces of attraction and repulsion between the particles and result in either dispersed or flocculated fabrics.

When the forces of repulsion dominate, that is in fresh water environments, the clay platelets align themselves to offer the maximum face to face area and thereby develop the maximum grain to grain distance (Fig. 5.6a). This produces strongly oriented fabrics which are dense, watertight and with void ratios which may be as low as 0.5. This type of fabric is character-

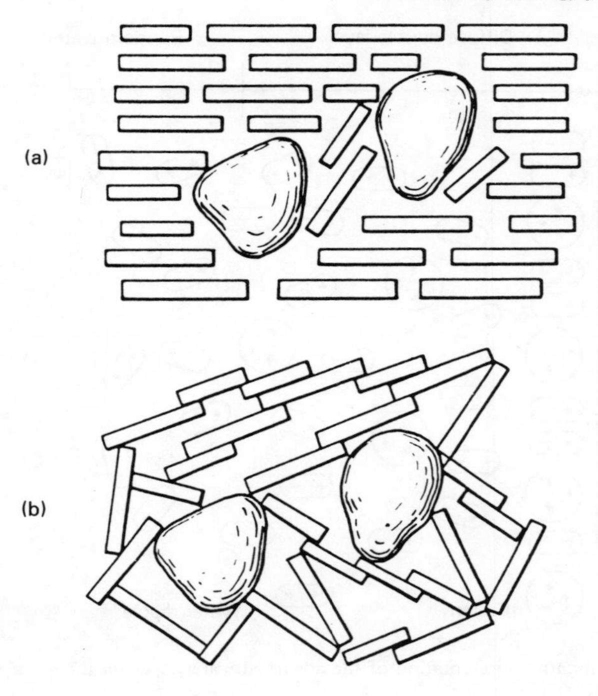

Fig. 5.6 (a) Non-flocculated and
(b) flocculated clay fabrics

istic of clays laid down in fresh water, formed by glacial action, or indeed compacted as fill under wet conditions during construction.

Where the forces of attraction are stronger than those of repulsion, then flocculent structures develop. Salt water is an extremely good electrolyte causing grains to collect together in a random way, trapping water within the large pore spaces between the particles (Fig. 5.6b). The particle to particle contacts may be edge to edge, edge to face or face to face (Fig. 5.7). Low densities and high compressibilities are characteristic of these soils. However, they tend to have relatively high strength and resistance to vibration because of the bound particles. They do though become very wet and sticky on remoulding, as the water in the voids is freed during the mixing and becomes added to the adsorbed layers. This marked softening on remoulding is called sensitivity and often presents problems on construction sites where remoulding by contractors plant can cause very difficult conditions even in dry weather.

In addition to the relationships between particles, it is now known that in many clays, aggregations or domains exist, comprising many individual clay mineral particles. The domains are made up of oriented clay particles, although the orientation between domains is usually random. The develop-

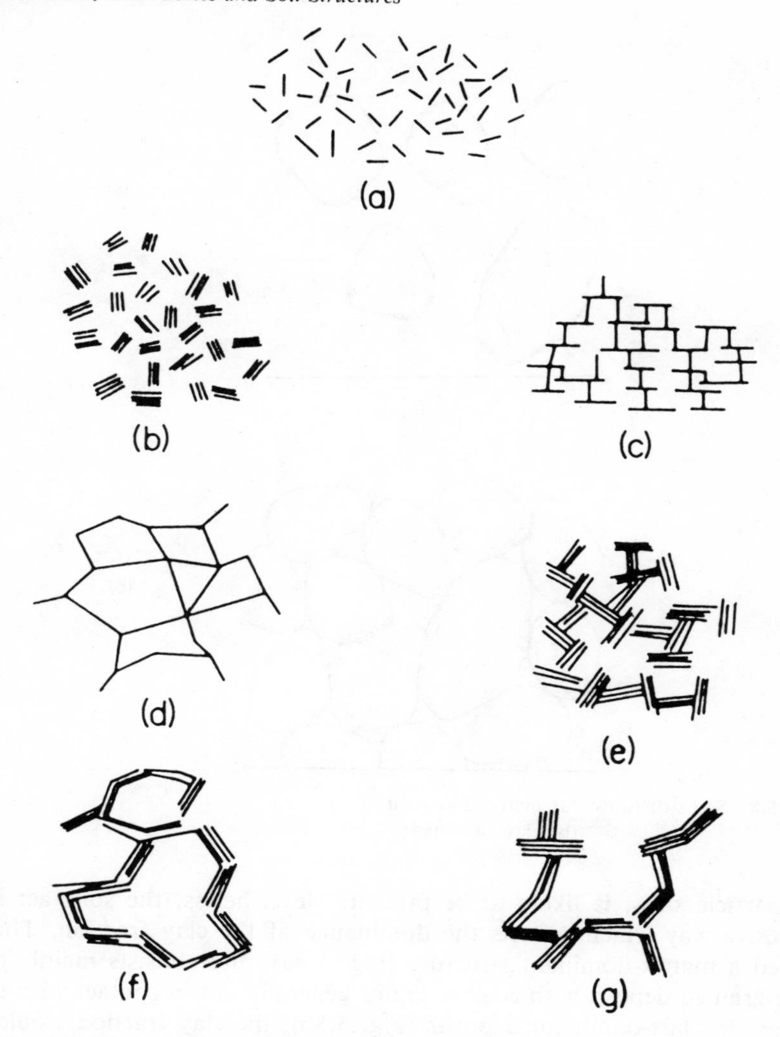

Fig. 5.7 Modes of particle associations in clay suspensions:
(a) dispersed — flocculated;
(b) aggregate — deflocculated;
(c) edge to face flocculated — dispersed;
(d) edge to edge flocculated — aggregated;
(g) edge to face and edge to edge flocculated — aggregated

ment of domains increases as the clay becomes consolidated, the domains gradually increasing in size and sometimes showing a general alignment with the bedding of the deposit. Aggregation of particles in this way is demonstrated in Fig. 5.7 (b, e, f and g).

Very few clay deposits consist entirely of clay minerals, and it is usual for clays to be silty or even sandy. In the case of some glacial clays, a full array

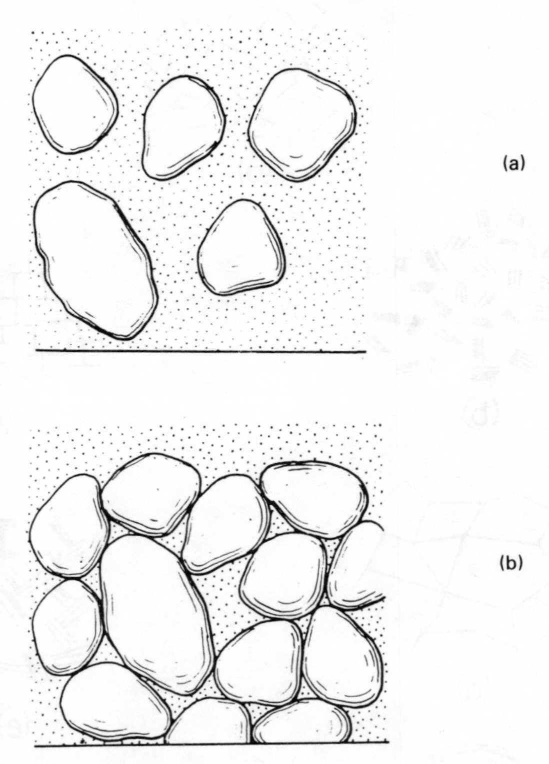

Fig. 5.8 Soil structures: (a) matrix dominant
(b) clast dominant

of particle sizes, is likely to be present. Nevertheless, the soils act in a cohesive way which reflects the dominance of the clay fraction. This is called a matrix-dominant structure (Fig. 5.8a), and consists mainly of a fine-grained deposit with coarser grains generally not in contact with each other. In clast-dominant deposits (Fig. 5.8b), the clay fraction would be generally ineffective and the soil would appear virtually cohesionless.

5.4 Soil Structures

The interrelationship between one bed of soil and another, that is the stratification, is an important element of soil structure. Post-depositional influences, especially the formation of discontinuities, also constitute a significant element of a soil mass. These two are the main parts of the soil structure.

Stratification is generally determined by the mode of deposition of sediment. Different bands of sediment in beds of differing shape, size, thickness, etc., will mean that properties within the soil mass vary from place to place. Careful examination of soil masses is necessary to determine the details of structure.

Homogeneous soils are relatively rare. Careful inspection usually reveals a certain non-uniformity of composition. In some of the major problem soils of South-East Asia, the transitional and marine clays of Singapore, bands of peat, sand, silt and shells occur in many locations (Pitts, 1983e, 1984a, 1984b). In the south of the island, fine sand and silt bands proved most influential in determining the settlement characteristics of the marine clay (Harvey, 1982). The rates of settlement determined by laboratory tests on samples of the clay which did not contain the stratification produced results which indicated a very slow rate of settlement and a very low permeability. Vertical drains installed to accelerate settlements by inducing better drainage were ineffective however, because drainage along the horizontal stratification was already of a high order. The rate of settlement proved much quicker than predicted because of the added facility for drainage afforded by the coarser layers.

Anisotropy of strength and permeability frequently results from stratification. Varves are seasonal bands of silt and clay found particularly in lake deposits in regions where freezing occurs in winter. One silt and one clay layer represent a year's deposition. These highly ordered sediments have very high horizontal permeabilities, but very low vertical ones.

The range of soil structures is of course enormous. It is important to be aware of likely sources of variability within a deposit however and its effect on the geotechnical behaviour. This can only be achieved if some understanding of the processes and products characteristic of a particular environment and mode of deposition are understood.

Discontinuities in soils usually result from shrinkage as the sediment compacts, stress relief as a result of unloading, rapid loading and tectonic deformation. The Old Alluvium of eastern Singapore is a braided river deposit comprising mainly sand and gravel (Gupta *et al.*, 1980). This formation has been extensively worked for construction and reclamation materials. The excavation walls in this material frequently show excellent sheet joints which may penetrate a metre or so into the soil mass. The sheet joints, formed parallel to the excavation surfaces, then have a marked influence on pit wall stability. In clays, fissures may be particularly important and have a profound influence on behaviour. The significance of fissures in clays forming cut slopes was mentioned in the section on stress relief. The effects on strength in general are well known, especially as they relate to laboratory samples. In some soil formations, faults have developed. These are commonly associated with indurated materials, but in soils too, they represent important discontinuities. They are often infilled with clay and occur in closer sets than is usually the case in rocks (Pitts, 1983c).

Not all features of soil fabrics and structure are natural and the broad subject of sample disturbance is a major concern when assessing the realism of laboratory test values on relatively small soil samples. Disturbance

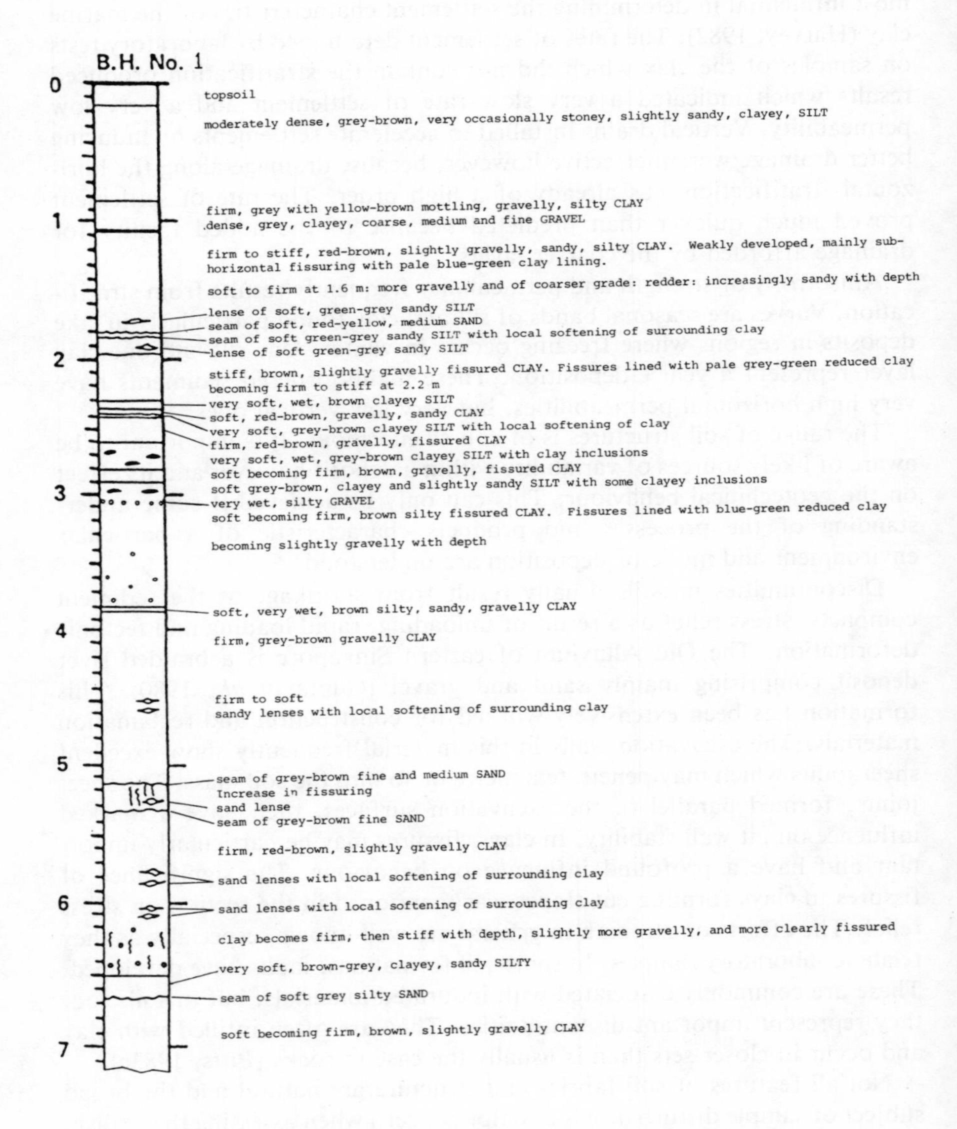
CLIENT: CALDY GOLF CLUB
JOB: SLOPE STABILIZATION
B.H. No. 1
0
1
2
3
4
5
6
7
topsoil
moderately dense, grey-brown, very occasionally stoney, slightly sandy, clayey, SILT
firm, grey with yellow-brown mottling, gravelly, silty CLAY
dense, grey, clayey, coarse, medium and fine GRAVEL
firm to stiff, red-brown, slightly gravelly, sandy, silty CLAY. Weakly developed, mainly sub-horizontal fissuring with pale blue-green clay lining.
soft to firm at 1.6 m: more gravelly and of coarser grade: redder: increasingly sandy with depth
lense of soft, green-grey sandy SILT
seam of soft, red-yellow, medium SAND
seam of soft green-grey sandy SILT with local softening of surrounding clay
lense of soft green-grey sandy SILT
stiff, brown, slightly gravelly fissured CLAY. Fissures lined with pale grey-green reduced clay
becoming firm to stiff at 2.2 m
very soft, wet, brown clayey SILT
soft, red-brown, gravelly, sandy CLAY
very soft, grey-brown clayey SILT with local softening of clay
firm, red-brown, gravelly, fissured CLAY
very soft, wet, grey-brown clayey SILT with clay inclusions
soft becoming firm, brown, gravelly, fissured CLAY
soft grey-brown, clayey and slightly sandy SILT with some clayey inclusions
very wet, silty GRAVEL
soft becoming firm, brown silty fissured CLAY. Fissures lined with blue-green reduced clay
becoming slightly gravelly with depth
soft, very wet, brown silty, sandy, gravelly CLAY
firm, grey-brown gravelly CLAY
firm to soft
sandy lenses with local softening of surrounding clay
seam of grey-brown fine and medium SAND
Increase in fissuring
sand lense
seam of grey-brown fine SAND
firm, red-brown, slightly gravelly CLAY
sand lenses with local softening of surrounding clay
sand lenses with local softening of surrounding clay
clay becomes firm, then stiff with depth, slightly more gravelly, and more clearly fissured
very soft, brown-grey, clayey, sandy SILTY
seam of soft grey silty SAND
soft becoming firm, brown, slightly gravelly CLAY

Fig. 5.9 An example of a logging sheet for a complex soil succession.

around the edges of samples close to the contact with the walls of sampling tubes is well known. Naturally occurring structures, for example fissures, may become more open as a result of stress relief when a soil is extruded from a tube. In some less common cases, total collapse of the sample has been known to occur during driving, and during reconstitution a new structure imparted (Pitts, 1983d).

5.5 The Description of Engineering Soils

In order to facilitate an accurate impression of soils in the mass, either description from exposures or from split undisturbed samples should be undertaken. Further descriptions based on index tests in the laboratory may be undertaken later, but only in these forms will the full relationships between the various soil layers present be appreciated.

Table 5.2 Some values of the common properties of soils

A. COHESIONLESS SOILS		
	Gravels	*Sands*
Specific gravity	2.5–2.8	2.6–2.7
Bulk density (Mg/m³)	1.45–2.3	1.4–2.15
Dry density (Mg/m³)	1.4–2.1	1.35–1.9
Porosity (%)	20–50	23–35
Shear strength (kPa)	200–600	100–400
Angle of friction	35–45°	32–42°
B. COHESIVE SOILS		
	Silts	*Clays*
Specific gravity	2.64–2.66	2.55–2.75
Bulk density (Mg/m³)	1.82–2.15	1.5–2.15
Dry density (Mg/m³)	1.45–1.95	1.2–1.75
Void ratio	0.35–0.85	0.42–0.96
Liquid limit (%)	24–35	Over 25
Plastic limit (%)	14–25	Over 20
Coefficient of consolidation (m²/yr)	12.2	5–20
Effective cohesion (kPa)	75	20–200
Effective angle of friction	32–36°	
C. ORGANIC SOILS AND FILL		
	Peat	*Coarse discard*
Moisture content (%)	650–1100	6–14
Specific gravity	1.3–1.7	1.8–2.7
Bulk density (Mg/m³)	0.91–1.05	1.2–2.4
Dry density (Mg/m³)	0.07–0.11	1.05–2.0
Void ratio	12.7–14.9	0.35–Over 1
Liquid limit (%)		23–45
Plastic limit (%)		Non-plastic – 35
Effective angle of friction	5°	28°–40°
Effective cohesion (kPa)	20	20–50

It is essential that the material, the fabric and the structure of the soil are described. Great care is required in order to include each change in soil type and represent fully the variability of the ground. An example from a succession of glacial deposits is shown in Fig. 5.9. In general, it requires that each layer should be described in terms of the consistency of the soil, that is soft, firm, stiff or hard, its colour, and the composition in as much detail as possible. In addition, irregularities or discontinuities within each layer, for example sand lenses in the example shown should be indicated. There are few problems likely to arise from a thorough examination of a soil mass, whereas a careless inspection may lead to considerable trouble. Soils are highly variable as the summary table of properties (Table 5.2) indicates. It is safest in general to assume that non-uniformity of conditions will be the rule rather than the exception so that a careful attitude will be adopted from the outset when faced with construction in or on a soil mass.

CHAPTER 6: GEOLOGICAL AND GEOTECHNICAL MAPS

6.1 Geological Maps and Strike Lines

Geological maps are records of the types and distribution of geological materials at or near to the surface of the earth. They are normally the result of the recording of outcrops of rock which occur over various parts of an area, followed by subsequent construction to enable the completed outcrop of pattern to be drawn. Outcrop maps can also be drawn from borehole information and geological sections constructed by using the same technique as for completing an outcrop map. The method used is that of strike lines or structural contours.

As the latter name implies, they are lines joining points of equal height. Unlike the normal contour lines which represent height on the ground surface, structural contours represent height on the top of a bed of rock, or indeed on any planar geological feature (Fig. 6.1). As contour lines join points of equal height on the ground surface, so do structural contours join

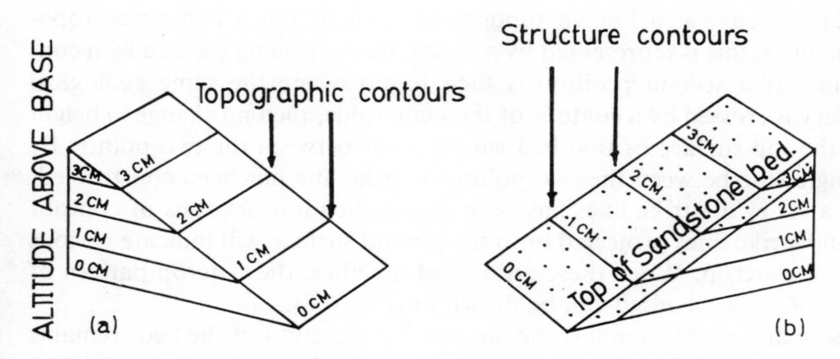

(a) Model 1, illustrating topographic contours on and inclined planar surface
(b) Model 1, illustrating structure contours on the of an inclined bed of sandstone.

Fig. 6.1 Structure contours on beds of rocks. (Open University, 1972; Copyright © 1972 by The Open University Press)

points of equal height on say, the top of a bed of rock. So, along any structural contour, there is no change of height, that is, zero gradient, and it therefore represents the direction of the strike of the bed (Fig. 3.1). Hence the name, strike line, (Fig. 6.2).

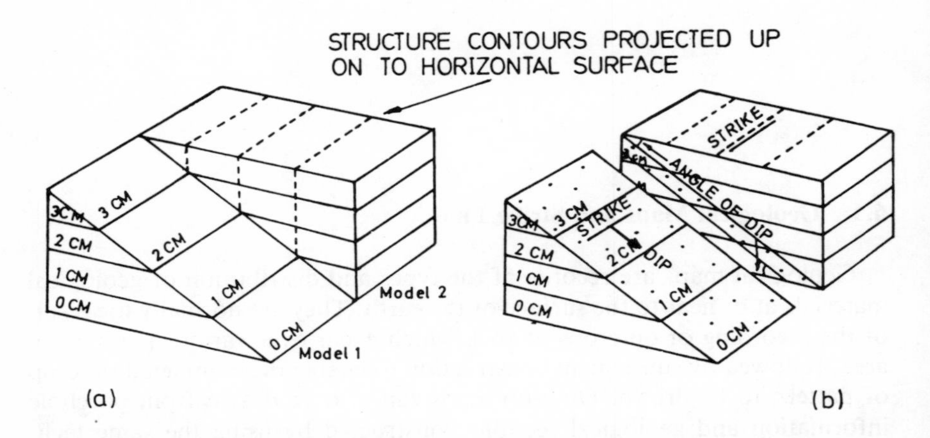

Fig. 6.2 Projection of structure contours from a rock bed to the ground surface. (Open University, 1972; Copyright © 1972 by The Open University Press)

The height of the top surface of a bed of rock is known only at certain points. The most obvious of these are where a surface outcrop of a rock boundary occurs at a known topographic height. On a contoured topographic map, this is represented by a rock boundary being crossed by a contour line. If a second position is then found where the same geological boundary is crossed by a contour of the same value, then no change in height along the top surface of that bed would occur between the two points. By drawing a line between the two points, a strike line has been constructed. When a series of strike lines has been drawn then intersections of contour lines and strike lines projected onto the ground surface will indicate various points of outcrop. When these are joined together, the outcrop pattern of the bed of rock recorded can be drawn in (Fig. 6.3).

With many problem maps, the dip and dip direction of the beds remains constant, and so as long as one strike line can be drawn through two intersection points of outcrop and contour, only one intersection point is required for other values. In fact, if any two strike lines can be drawn and it is assumed that the dip is constant, then the spacing between strike lines of any height interval can be calculated and the series of strike lines drawn in

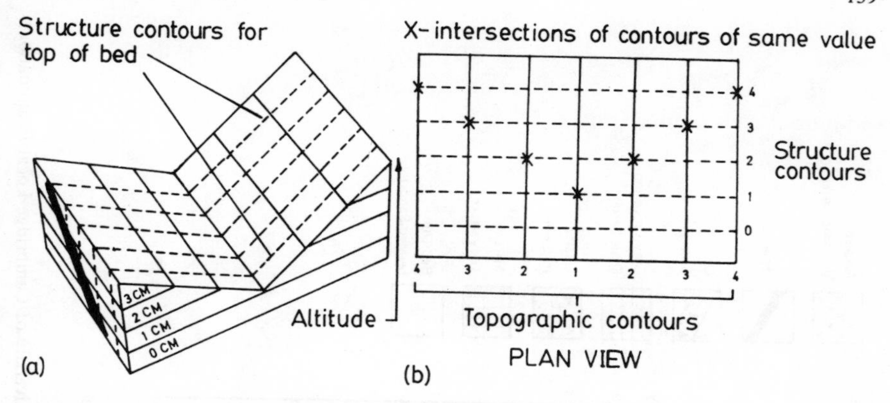

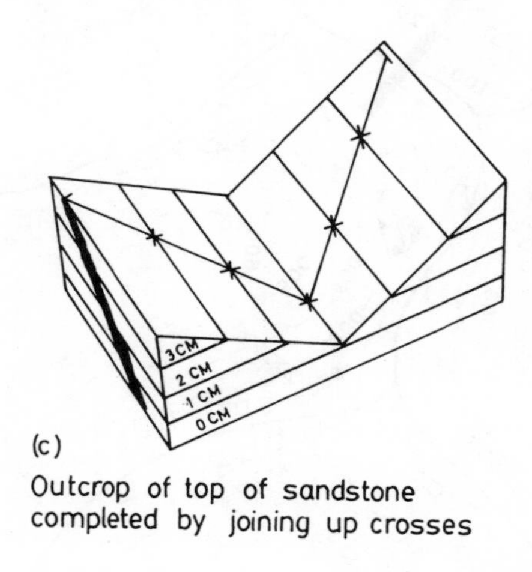

Fig. 6.3 Use of structure contours to plot the outcrop pattern of a bed of rock. (Open University, 1972; Copyright © 1972 by The Open University Press)

parallel to the first one constructed, and at an even spacing. Dip of a bed is only a gradient, and if the rate of height change does not vary, then neither does the gradient or dip of the bed.

In Fig. 6.4a a partially completed outcrop map is presented with a variety of geological features recorded on it. The map requires the outcrop pattern to be completed. In this case, dips and dip directions of any set of rocks are considered to be constant.

A number of starting points offer themselves, but the completion of the outcrops of the linear features will be undertaken first. Straight lines are a rarity in nature, and in geology, such an occurrence usually indicates a vertical feature. The legend of Fig. 6.4a indicates that the linear features

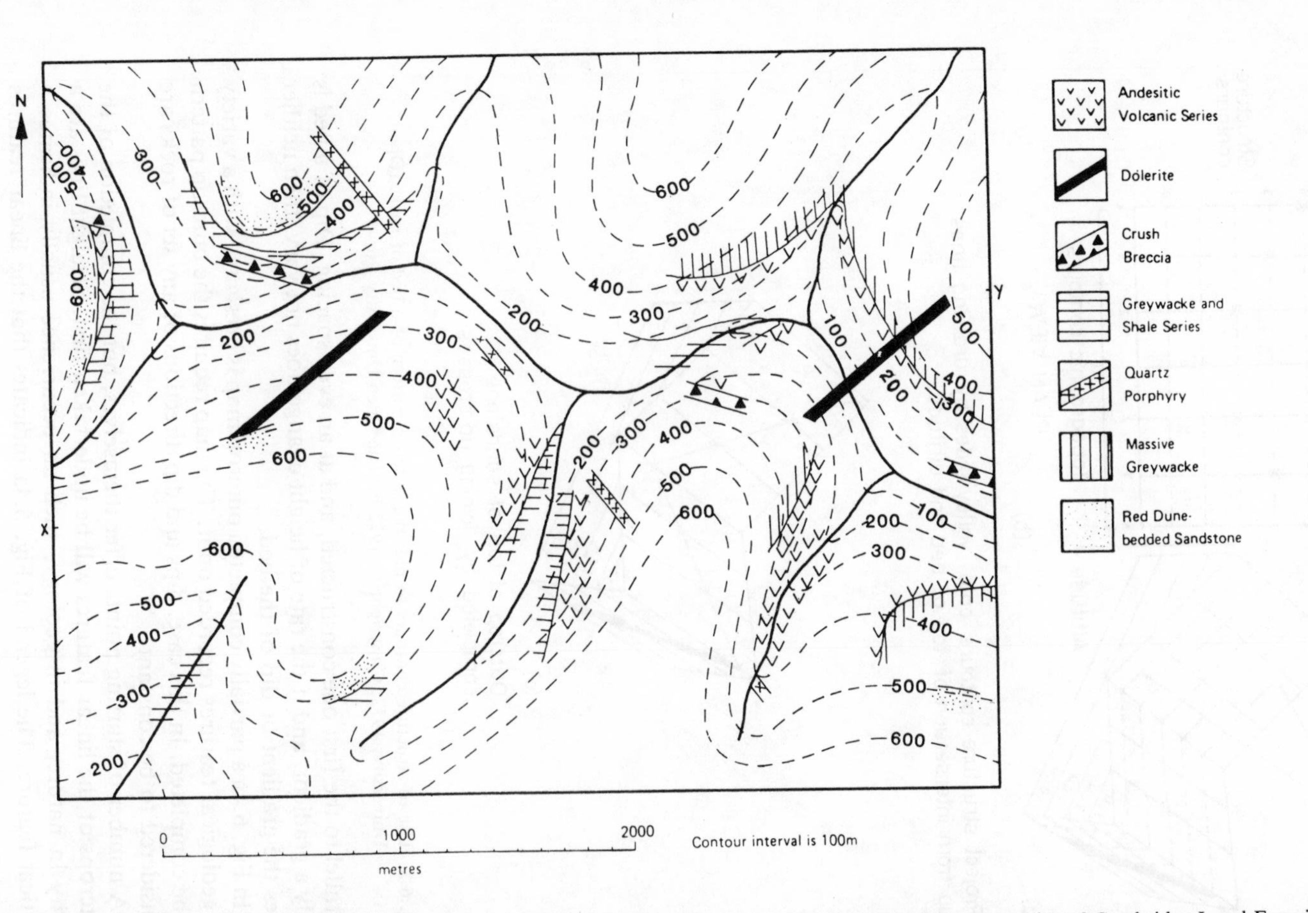

Fig. 6.4 (a) Partially completed geological map showing positions of recorded outcrops. (Modified after University of Cambridge Local Examinations Syndicate, 1979)

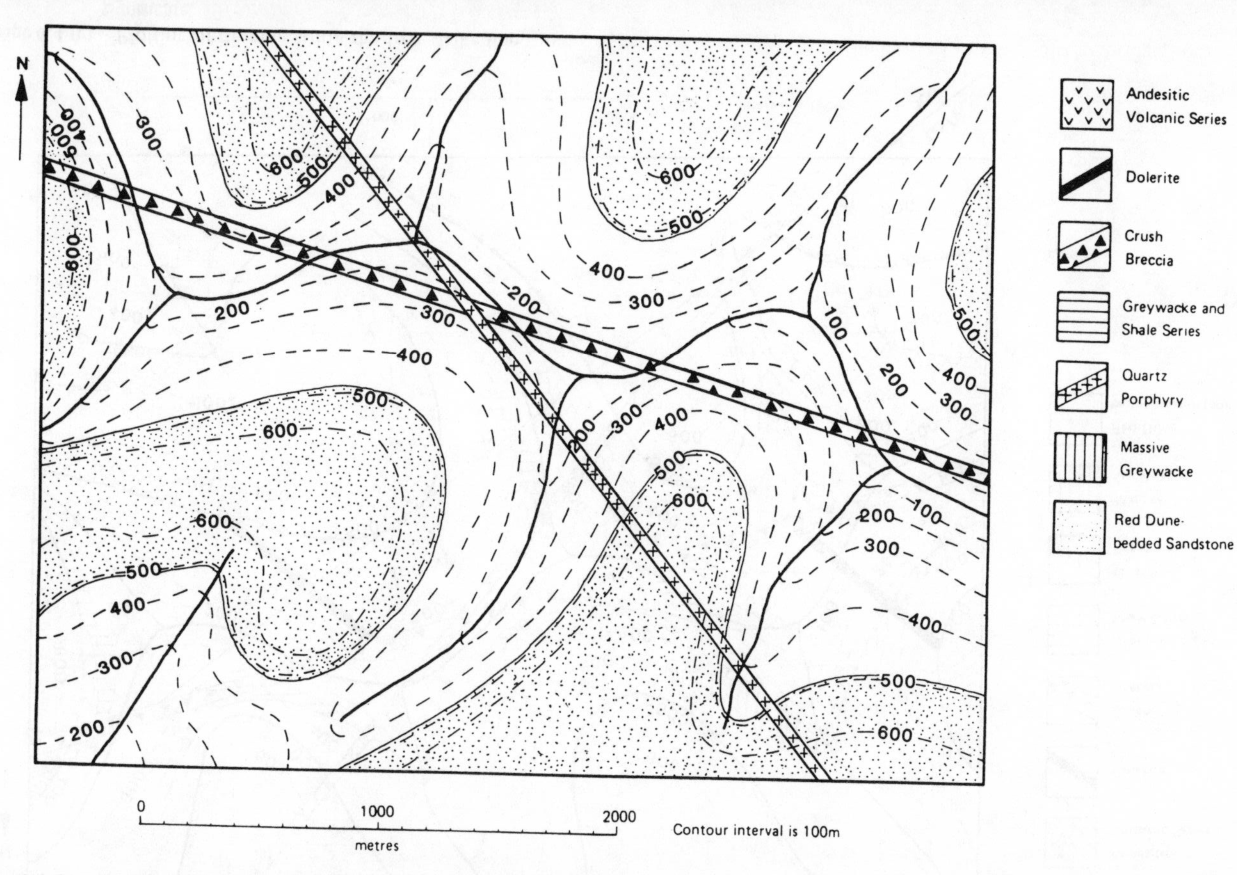

Fig. 6.4 (b) Plotting of linear geological features and horizontal beds.

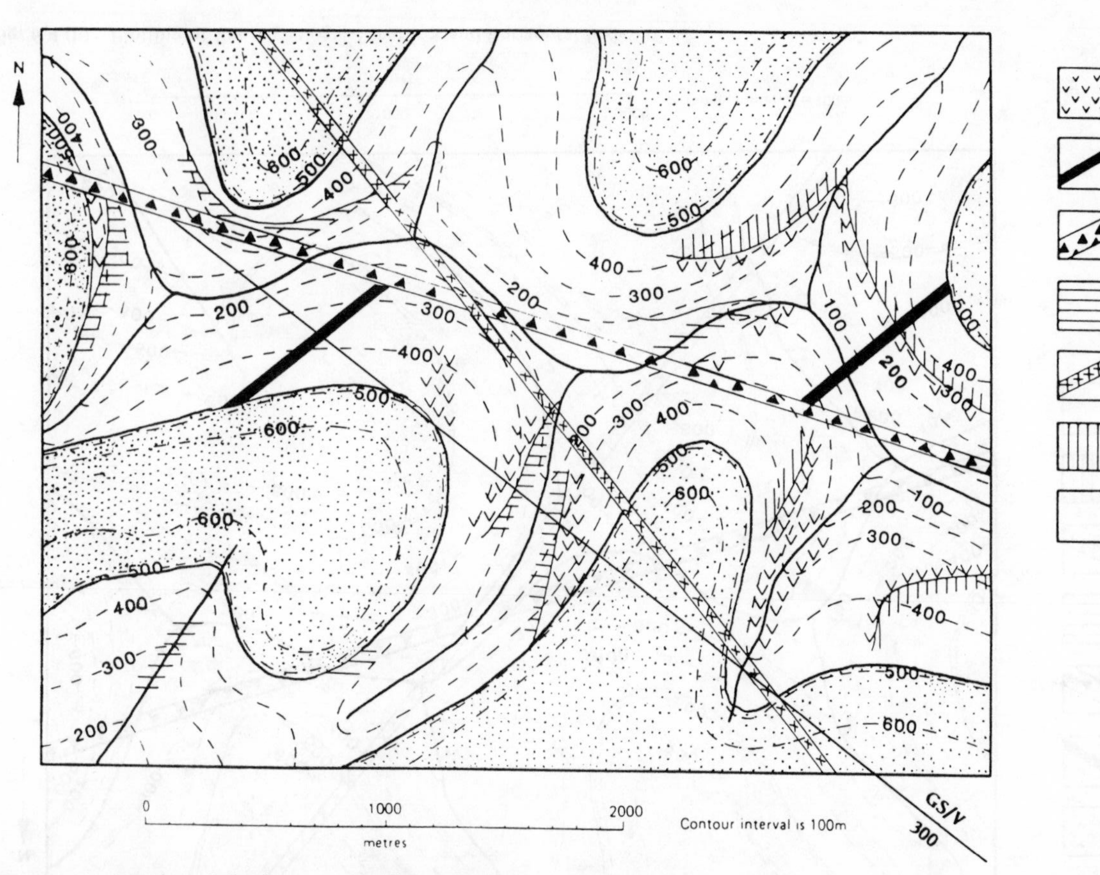

Fig. 6.4 (c) Plotting of dipping strata the first strike line.

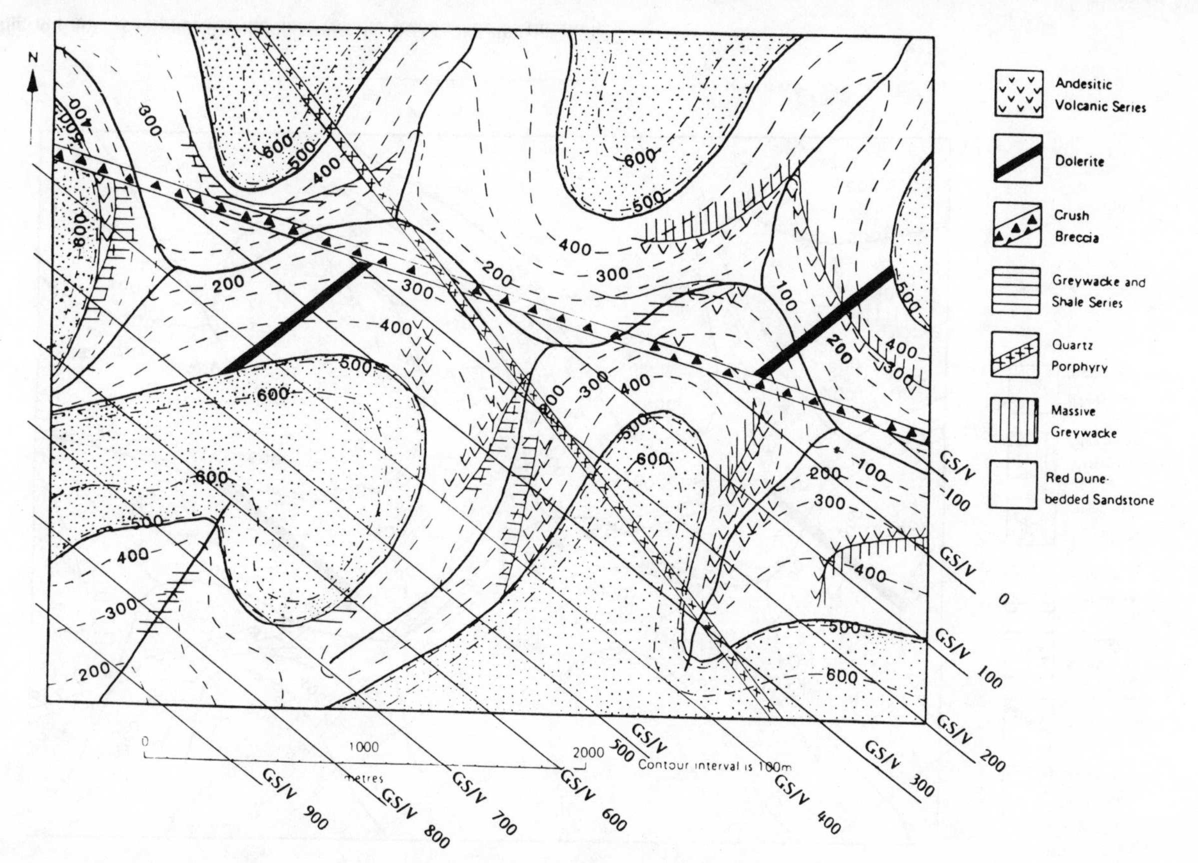

Fig. 6.4 (d) Completing the drawing of strike lines on one side of the fault.

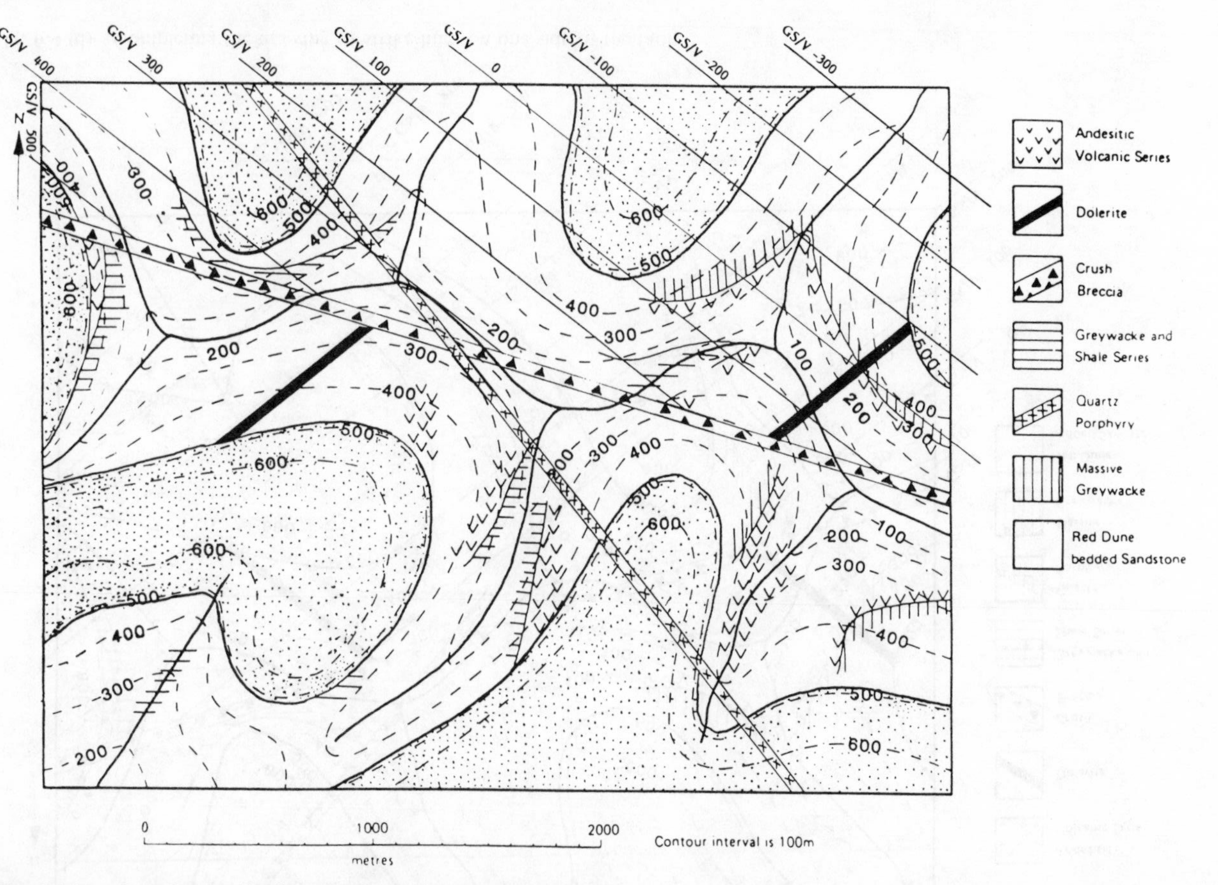

Fig. 6.4 (e) Completed strike lines for the second side of the fault.

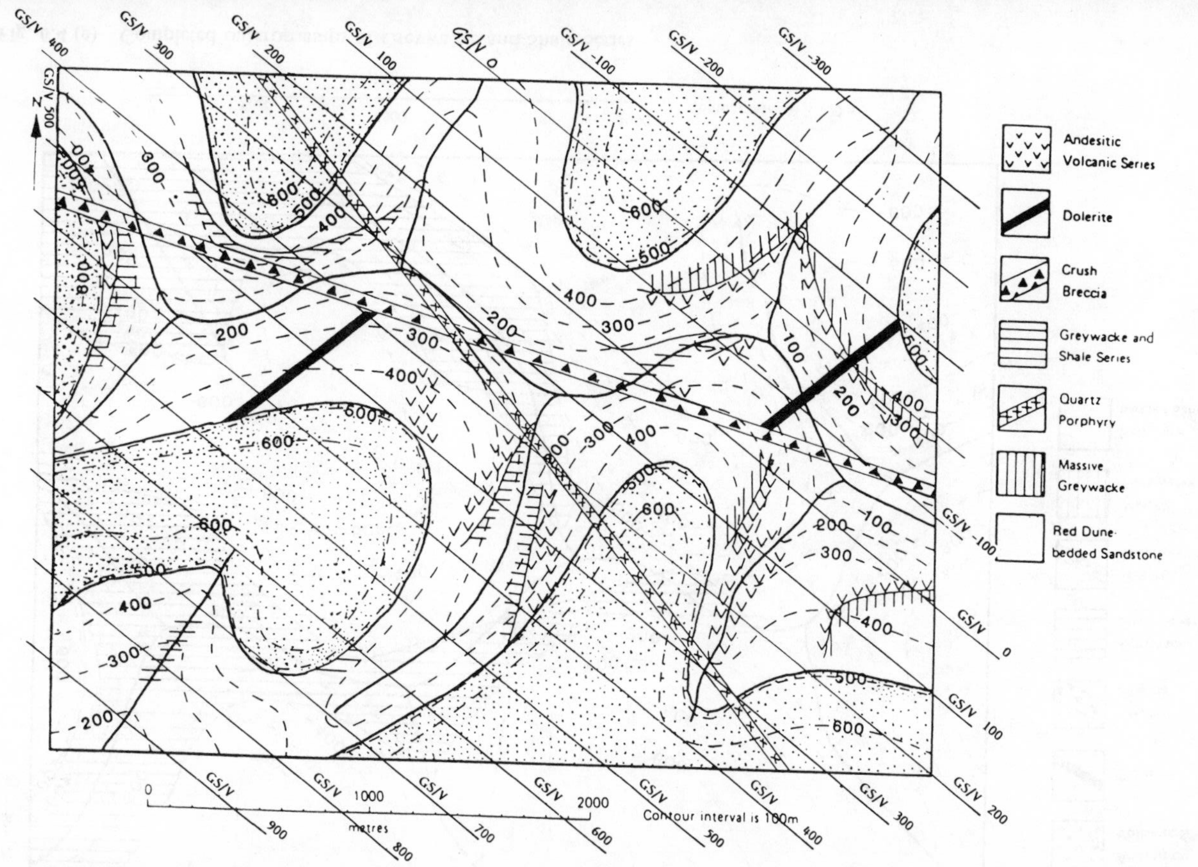

Fig. 6.4 (f) Completed strike lines for Greywacke and Shale Series Andesitic Volcanic series boundary.

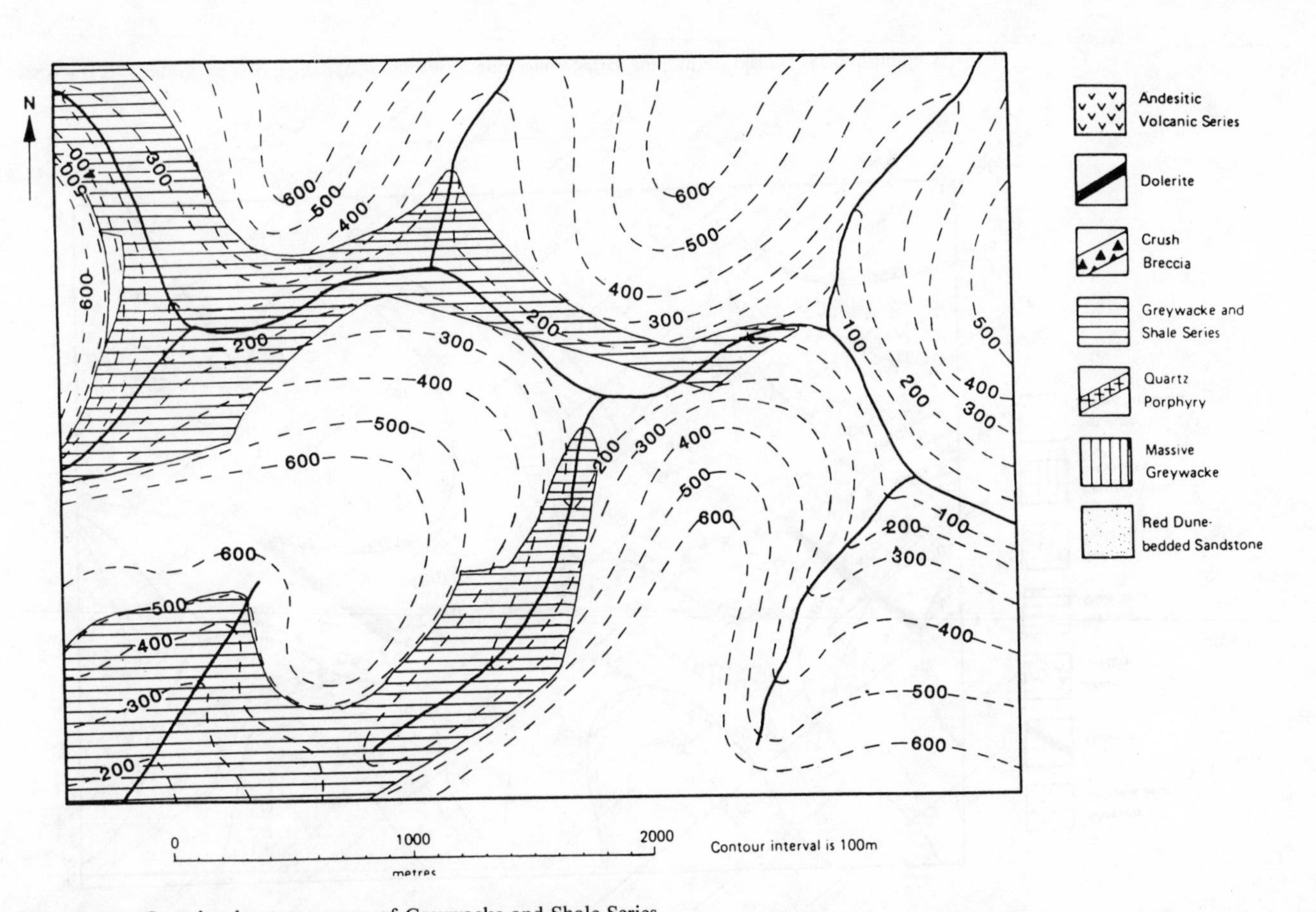

Fig. 6.4 (g) Completed outcrop maps of Greywacke and Shale Series.

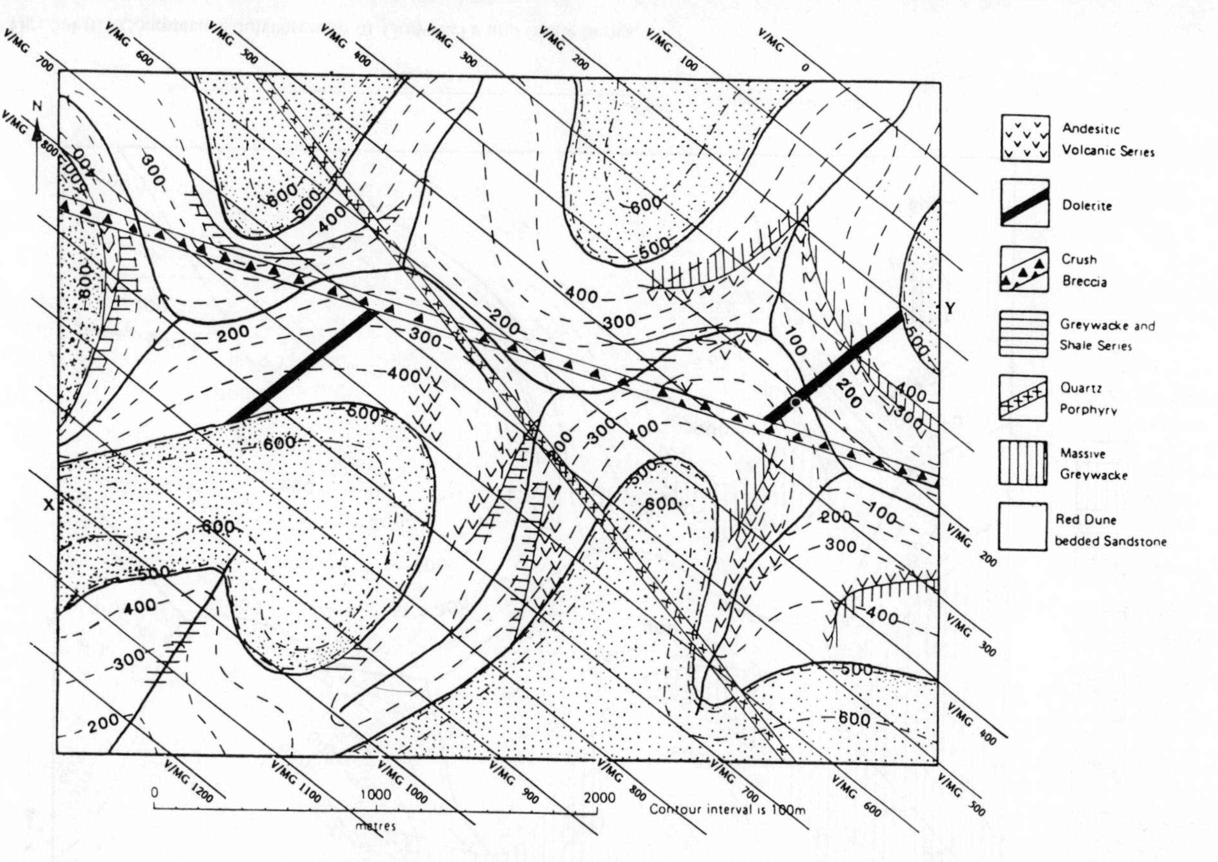

Fig. 6.4 (h) Completed strike lines for boundary between Andesitic Volcanic Series and Massive Greywacke.

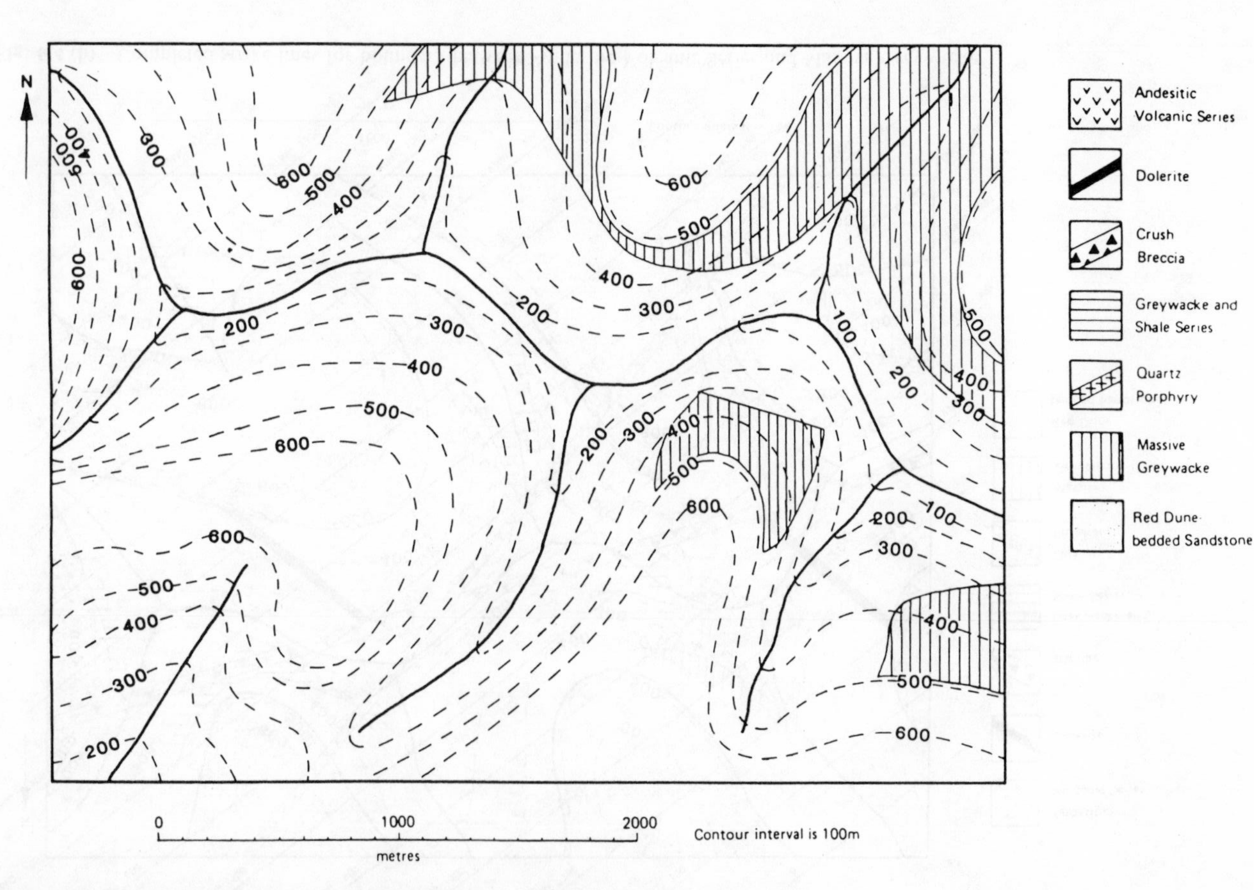

Fig. 6.4 (i) Completed outcrop map of Greywacke and Shale Series.

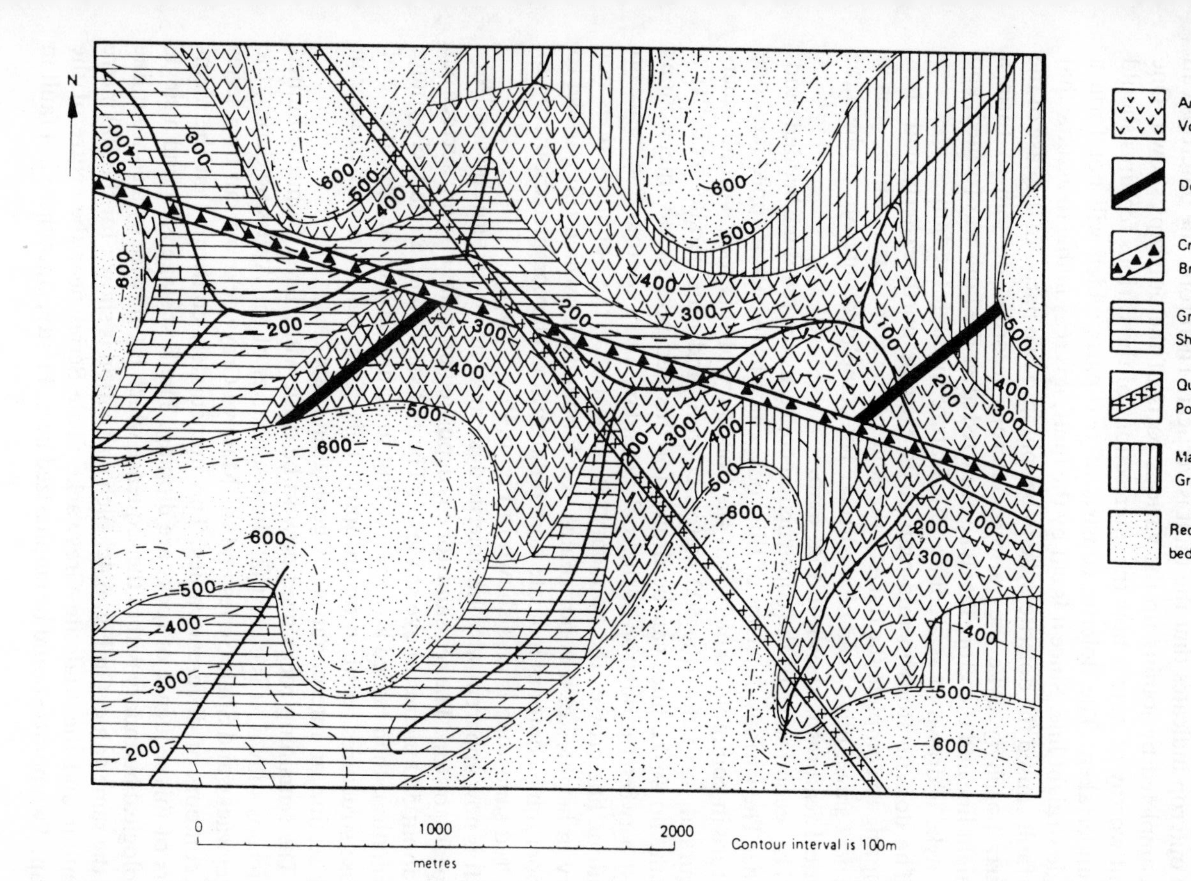

Fig. 6.4 (j) Completed geological map.

consist of dolerite, quartz porphyry and crush breccia. The first two are medium-grained igneous rocks commonly found in small scale intrusions. The outcrops found indicate that their form is vertical, and hence it is reasonable to assume that they are dykes, (Fig. 2.3). Crush breccia is an unusual material formed from (usually) coarse, ill-sorted angular fragments. The most common situation in which materials of this sort are found is along a fault, where the crush breccia constitutes the gouge. The straight line outcrop indicates that the fault is vertical. Therefore, the outcrops can be completed by joining up the various parts of the outcrops shown. One point worthy of note is how the intersections of the various features should be undertaken. The dolerite consists of two parts which will not form a single straight line. Since it is cut by the fault, it is reasonable to assume that the fault caused the dolerite dyke to be split into two parts. In the case of the quartz porphyry dyke the various outcrops recorded do join to form a straight line and therefore the dyke occurred after the fault was formed, and the dyke is drawn in as a continuous feature through the fault.

The dolerite stops in mid-map at its extreme north–easterly end. It is tempting to complete the dyke to the edge of the map, but as will be seen in the next stage, this would be incorrect. The completed outcrops of the three vertical features is shown in Fig. 6.4b.

The second stage of the construction concerns the red dune bedded sandstone. The various outcrops of this rock which have been mapped have one feature in common, namely that the boundary of this rock with all others occurs at just below the 500 metres contour and runs parallel to it. As a result, no strike lines can be drawn for this bed because no intersections exist between the outcrop of the rock boundary and a contour line. Since no contour line crosses the boundary, it is clear that the boundary does not vary in height, i.e. it is horizontal. Rock boundaries which run parallel to topographic contours indicate horizontal strata. The completed outcrop of the red sandstone is also shown in Fig. 6.4b.

It is now apparent why the dolerite dyke at its north-eastern end suddenly appears to stop. The red sandstone is deposited on a flat erosion surface cut at about 500 metres above sea level. The red sandstone, a relatively young formation, then buried any pre-existing structures giving the appearance of those structures terminating abruptly. This effect will be discussed later in the section on interpretation of geological maps.

The remaining rocks need now to be dealt with. Even from the limited outcrops shown, it is clear that the boundaries of the Andesitic Series, the Greywacke and Shale Series and the Massive Greywacke are cut by a series of contours of different values. This immediately confirms that the members of this sequence of rocks are dipping. The first stage is to find a single geological boundary within the sequence which is cut twice by a contour line of the same value. On Fig. 6.4c, such a situation is shown for the 300 metre contour and the Andesite–Greywacke Shale Series near the centre of the map. Two points should be emphasized here. Firstly, since there is a fault in this area, the strike line should run up to the fault but not cross it. Faults displace beds and thus change their height. The strike line for a bed on either side of a fault, even if it could be drawn as a continuous line through the fault, would not have the same height value. The strike line would only be a continuous line through the fault if the throw of the fault produced a vertical change in height of the same magnitude as or multiple of the strike line increment. The second point is that *all* strike lines should be properly labelled. In the case of the example shown (Fig. 6.4c), V stands for Andesitic Volcanic Series, GS for Greywacke Shale, and the height of the strike line is also given.

Additional strike lines for this rock boundary can now be drawn in parallel to the first one through each contour–rock boundary intersection. The strike lines may be completed by drawing further ones at an equal spacing to those already established (Fig. 6.4d). This is acceptable in this example as, it is assumed that the dip and strike of the tilted strata remain constant over the area of the map. Strike lines can be drawn to the north of the fault (Fig. 6.4e) producing a final pattern of strike lines for this one stratigraphic boundary as shown in Fig. 6.4f.

The outcrop of the boundary between the Andesitic Volcanic Series and the Greywacke and Shale Series can now be completed. Each intersection point of a strike line of a particular height and contour line of the same height indicates a point of outcrop. When all such intersections have been found, they can be joined-up to form the outcrop pattern, shown in Fig. 6.4g.

This procedure is now repeated for the boundary of the Andesitic Volcanic Series and the Massive Greywacke. The labelled strike lines for the Andesite–Massive Greywacke boundary and the completed outcrop are shown in Fig. 6.4h and i. The completed outcrop map without construction lines is shown in Fig. 6.4j.

Although several simplifications have been introduced in this example, the principle holds good for far more complex cases. If strikes and dips are not uniform, then by constructing strike lines strictly, this fact will emerge, and strike lines would be neither parallel nor evenly spaced.

Now that a full outcrop pattern has been established, it is possible to draw geological sections from the map. On the map in question, a geological section along the line X–Y is to be constructed across the map. The first task though is to draw the topographic profile along this line.

On the geological map, the edge of a plain piece of paper is placed along the line X–Y. The positions of X and Y are marked, as are the positions and values of contour lines as they are intersected by the edge of the paper (Fig. 6.5). The plain paper with contour details is then transferred to a piece of graph paper with a base line and vertical axis marked with appropriate elevations. The contour details are marked off on the graph paper at appropriate levels (Fig. 6.6a), and the points joined-up to form the profile (Fig. 6.6b).

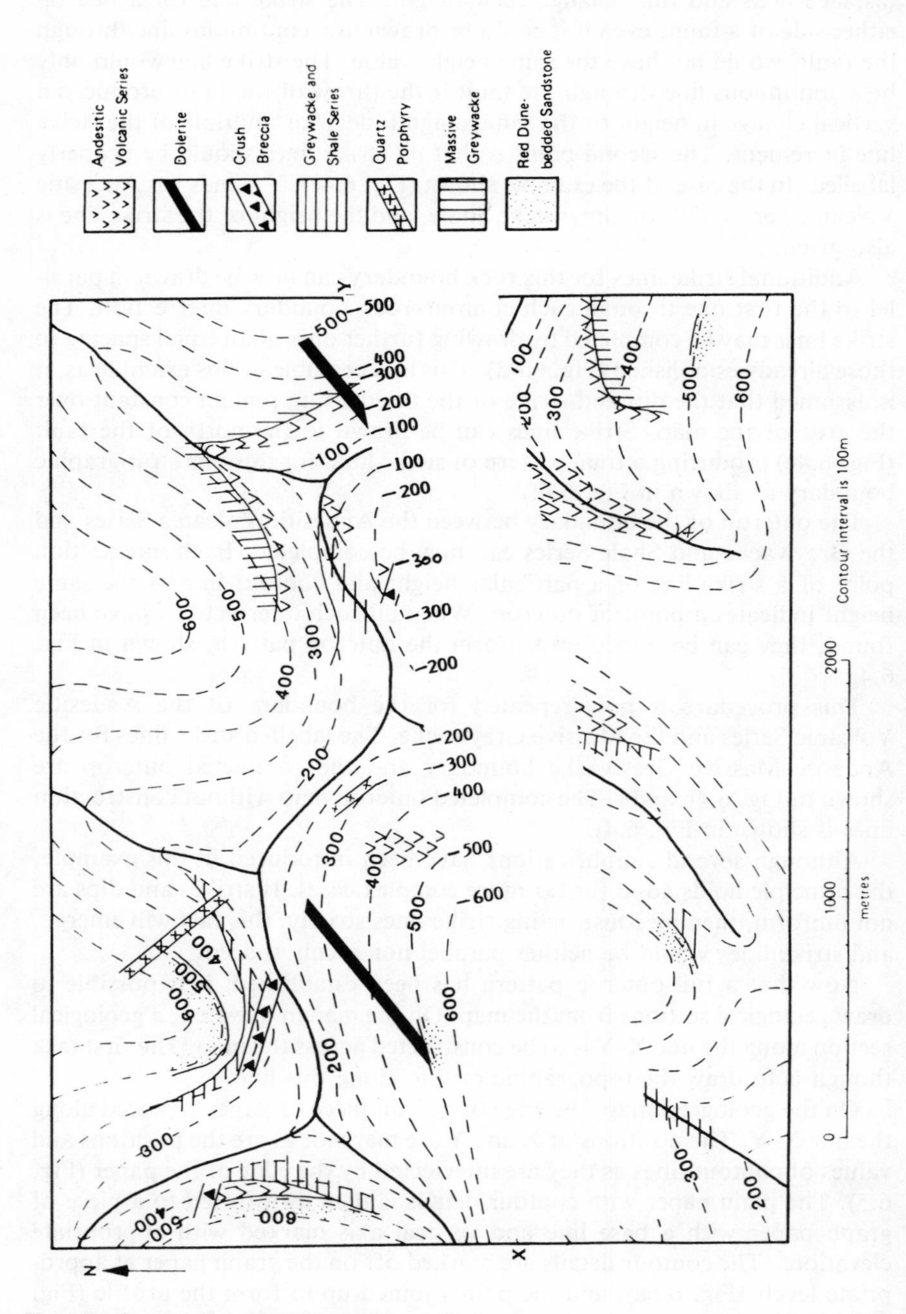

Fig. 6.5 Construction of a topographic profile.

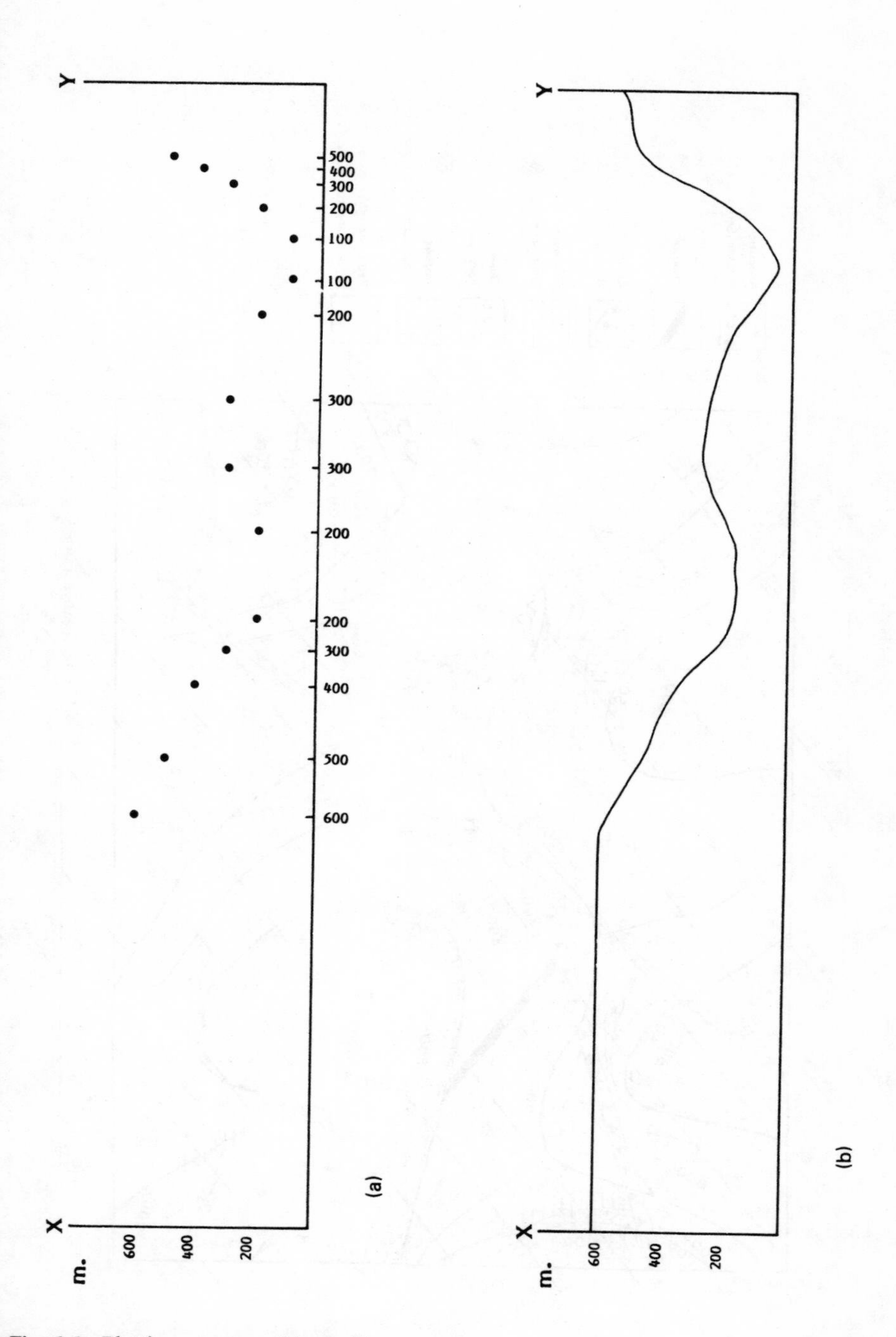

Fig. 6.6 Plotting a topographic profile.

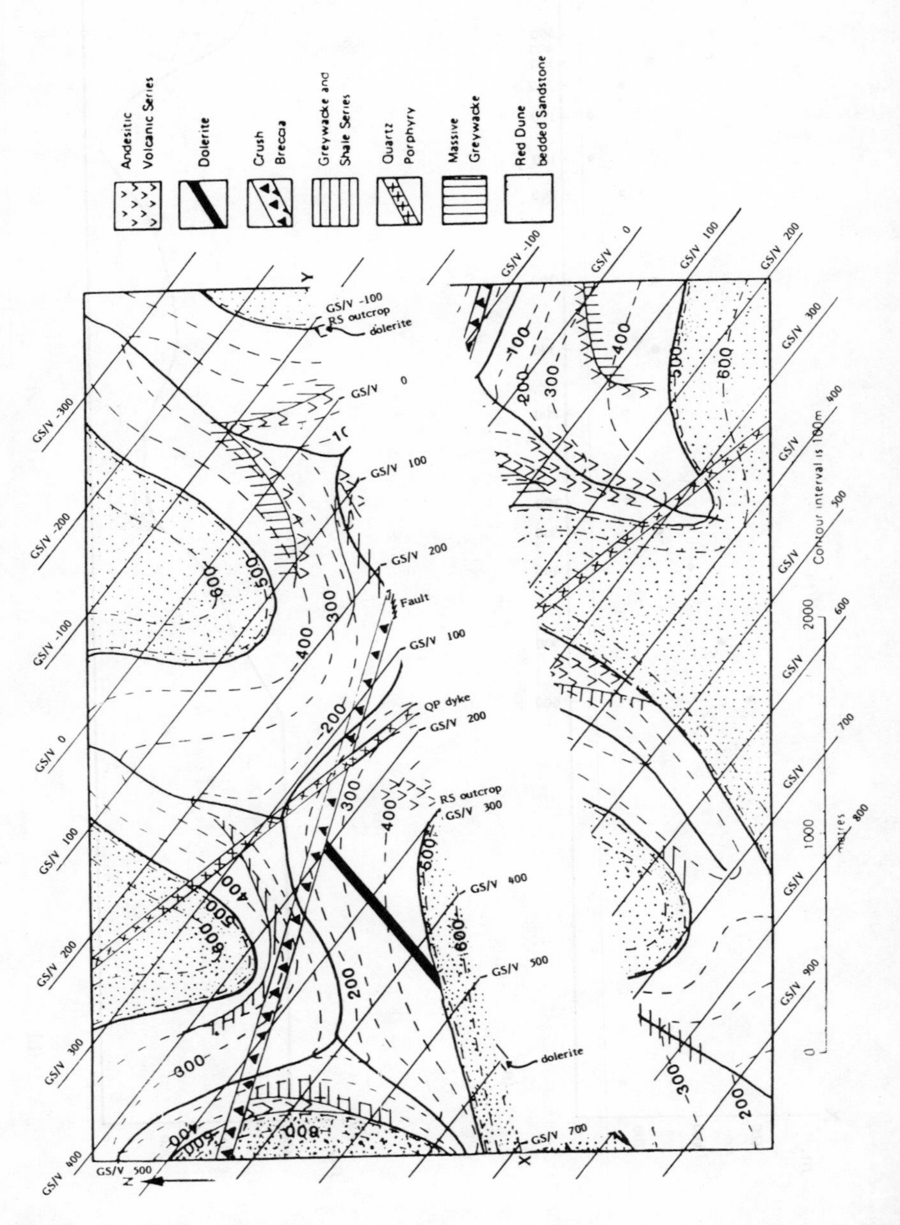

Fig. 6.7 Constructing a geological cross section.

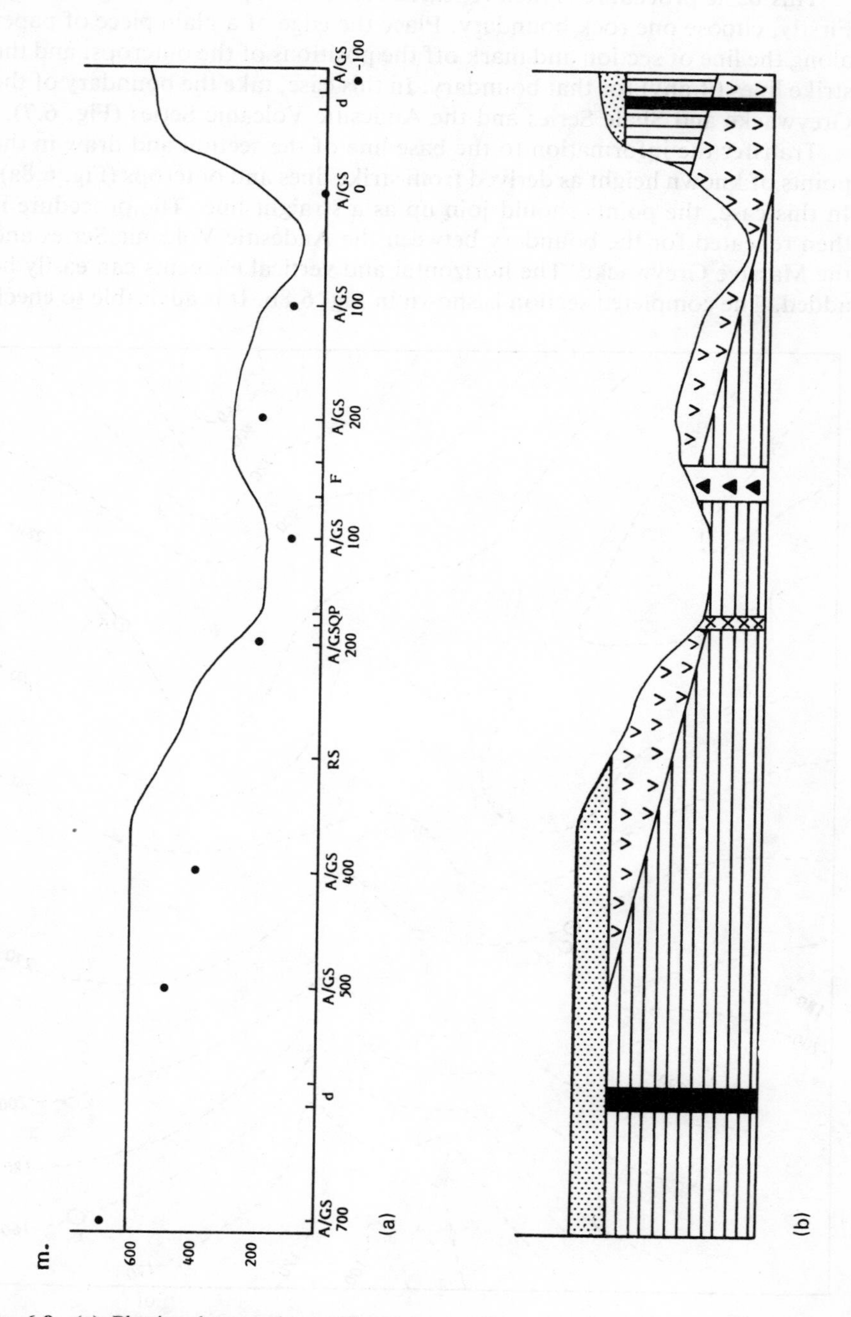

Fig. 6.8 (a) Plotting data on the profile, and
(b) completing the geological cross section.

This basic procedure is then repeated for the completed geological map. Firstly, choose one rock boundary. Place the edge of a plain piece of paper along the line of section and mark off the positions of the outcrops, and the strike lines (if any) for that boundary. In this case, take the boundary of the Greywacke and Shale Series and the Andesitic Volcanic Series (Fig. 6.7).

Transfer the information to the base line of the section and draw in the points of known height as derived from strike lines and outcrops (Fig. 6.8a). In this case, the points should join up as a straight line. The procedure is then repeated for the boundary between the Andesitic Volcanic Series and the Massive Greywacke. The horizontal and vertical elements can easily be added. The completed section is shown in Fig. 6.8b. It is advisable to check

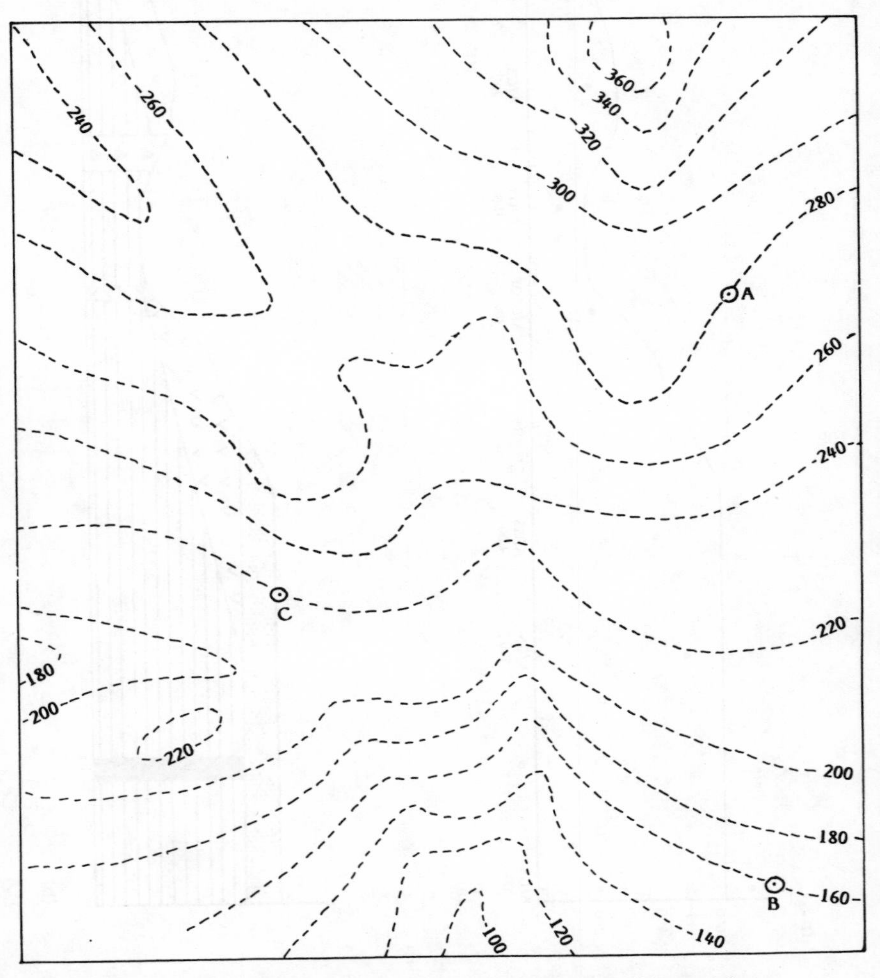

Fig. 6.9 (a) Three Point Problems: Contour map showing the positions of three boreholes.

that no serious errors have been made by following the order of outcrop along the line X–Y on the map, and ensuring that the same order occurs on the cross section.

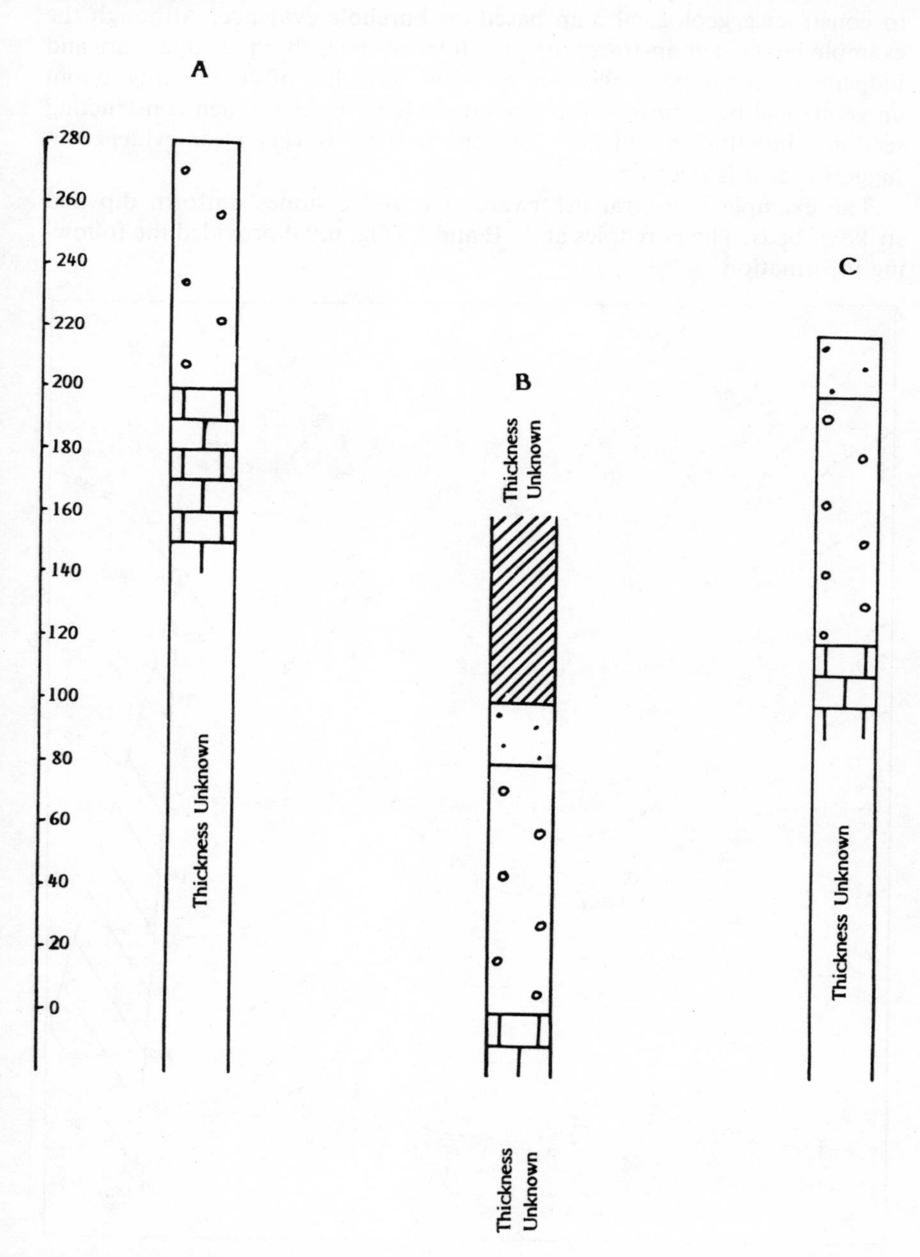

Fig. 6.9 (b) Geological logs of data from the three boreholes.

6.2 Three Point Problems

If no clear outcrop pattern exists in a particular area, it may be necessary to construct a geological map based on borehole evidence. Although the example below demonstrates the principles of the technique, great care and judgement need to be exercised in real situations. It is often tempting to join up geological boundaries from one borehole to another when constructing sections, but this should only be done if there is very clear evidence to suggest that it is realistic.

The example is a straightforward one and assumes uniform dip and strike of beds. The boreholes at A, B and C (Fig. 6.9a) provided the following information.

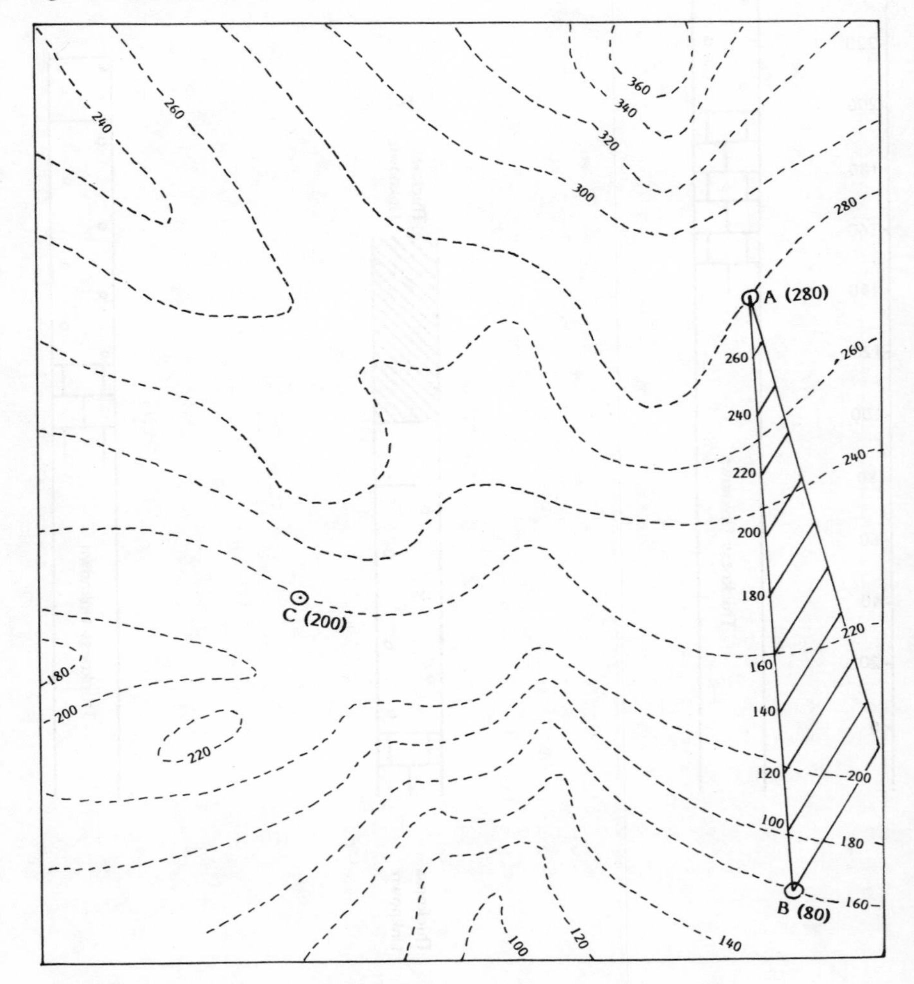

Fig. 6.9 (c) First stage of construction of the outcrop map.

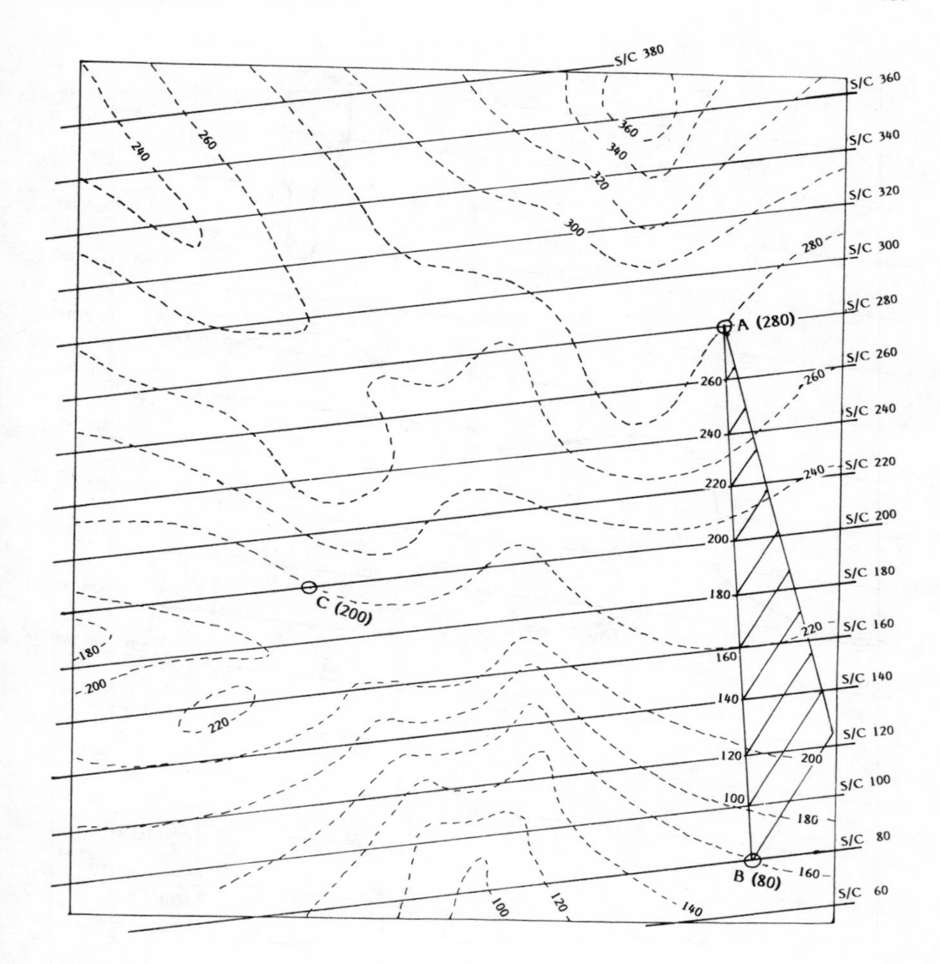

Fig. 6.9 (d) Construction of strike lines.

At borehole B, the succession was; clay, 60 metres, sandstone 20 metres, conglomerate 80 metres, limestone 70 metres. At C, the base of the clay appears, and at A, the base of the sandstone appears.

This information is best drawn in the form of borehole logs as shown in Fig. 6.9b. The first stage of the construction is to determine the nearest approximation to the direction of the dip from the information provided, and hence, via construction, determine the position of the first strike line. This direction will not, except in the most fortuitous of circumstances be the true dip. This is found by choosing one particular boundary and finding the maximum height of occurrence of that boundary. In this case, the sandstone–conglomerate boundary has been chosen. The choice is quite

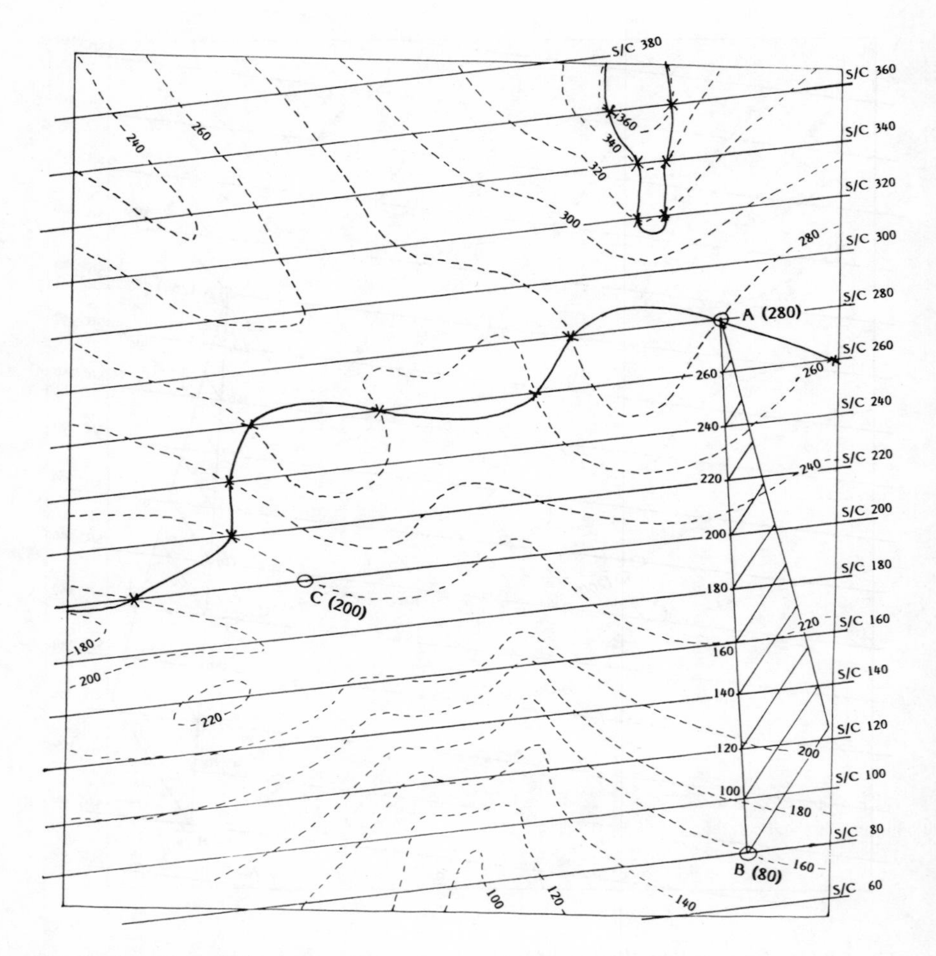

Fig. 6.9 (e) Drawing in the outcrop of a rock bed.

arbitrary since even the level of a now eroded formation, for example the clay–sandstone boundary in borehole A, could be determined and utilized. The top of the conglomerate occurs at 280 metres in borehole A, 80 metres in borehole B, and 200 metres in borehole C. The greatest height difference over what are similar horizontal distances, provides the largest gradient, from the given information. In this case, it occurs between boreholes A and B, and a line may be drawn between these two points.

A strike line joins points of equal height on a bed and runs perpendicular to the direction of the true dip. The height of the chosen sandstone–conglomerate boundary in borehole C occurs at 200 metres, that is, 20 metres from the top of the borehole. A strike line would then pass through this

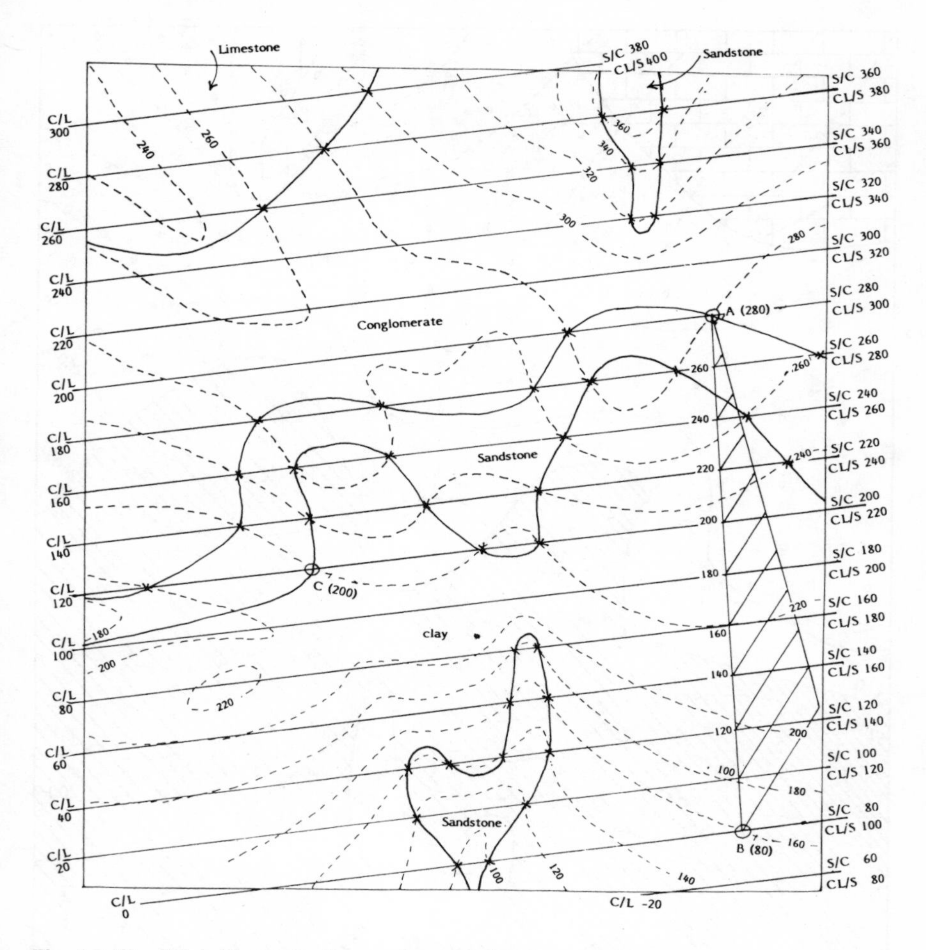

Fig. 6.9 (f) Relabelling strike lines and completing the outcrop pattern.

point to a position on the line AB which is known to be at 200 metres above sea level. At this point, it is worth a reminder that the line AB, and the proposed line from C to a point on AB are not at the surface of the ground, but on the surface of the conglomerate buried beneath differing amounts of overburden.

Since the dip of the beds is assumed to be constant, by dividing the line AB into an equal number of parts between 280 metres (the height of the boundary at A) and 80 metres (the height at B), the position where the bed is at 200 metres can be determined. This construction is shown in Fig. 6.9c.

Once the 200 metre position on AB is found, a line from that point to borehole C can be drawn. The line joins two points of equal height on top of the conglomerate, and this is therefore a strike line. Further strike lines at

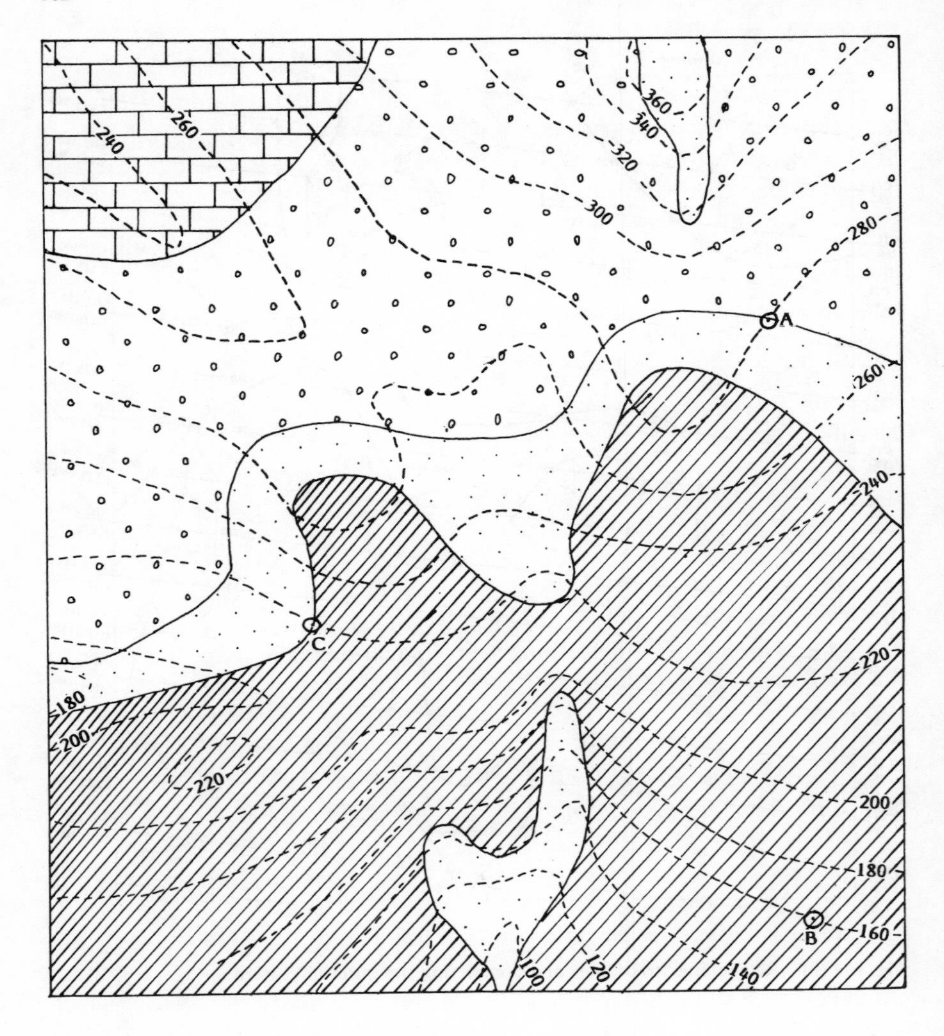

Fig. 6.9 (g) The completed outcrop map.

20 metre intervals can then be drawn parallel to this, and labelled accordingly, (Fig. 6.9d). The intersection points of the strike lines and contour lines of the same value can then be marked and joined-up to show the outcrop pattern of that boundary, (Fig. 6.9e). In the example, the beds all have thicknesses in increments of 20 metres. The height interval of the strike lines is also 20 metres and therefore the set of strike lines constructed will be the same for all the beds. Of course, the strike lines will need to be re-labelled for each bed in turn as shown in Fig. 6.9f. The outcrop pattern for each bed can be drawn in turn as shown in Fig. 6.9f. The final map is shown in Fig. 6.9g.

6.3 Interpretation of Geological Maps

For most purposes of learning about geological maps and their interpretation, problem maps printed in black and white and using symbols, are normally the starting point. Using such maps, particular geological features may be demonstrated in a classic form without the inevitable complications found on actual geological maps.

Most countries of the western world have very extensive geological coverage of their lands, and developing countries are tending to apply the long-established principles employed in those countries to their own programmes of geological mapping.

The most striking feature of geological maps is their colour. Usually, these are determined for sedimentary rocks by their geological age, and certainly for the Phanerozoic time scale are usually standardized. Unfortunately, the standardization is within one country, and there are no agreed standards of colour between countries. Hence, it is essential that care is taken when changing from the maps of one country to those of another.

The outcrop shapes denoting the various geological structures are the same on black and white maps as they are on coloured ones. Therefore, the principles of interpretation do not vary and simplified problem maps are a good way to learn. In describing a geological problem map, the order of the succession and its configuration are usually taken as the starting point. It is a good idea when interpreting geological maps to try to find the oldest rock first. This can be done in a number of ways, and is quite possible even if such details are not given in the key to the map, for example:

(a) work down the succession by observing the dip arrows or other symbols of dip and strike, (Fig. 6.10);
(b) observe the behaviour of beds in valleys; with dipping strata; the outcrop will usually "V" in the direction of dip, (Fig. 6.11);
(c) observe the course of outcrops in relation to topographic height, but be careful, as synclines do not always produce valleys, or anticlines always produce mountains; horizontal beds will have boundaries which run parallel to topographic contours;
(d) observe the relationships between igneous and sedimentary rocks, and try to determine the time relationships.

The structure of the rock requires interpretation. Some geological maps provide at least a partial interpretation of the folding, although quite often it is of minor rather than major folds. By interpreting the nature of folding information pertinent to the interpretation of the rock succession will also be provided. The oldest beds in a folded succession are likely to occur at the core of an anticline, and the youngest of that succession at the core of a syncline (Fig. 6.12). The main features of rock structure to be recorded are:

(a) the type, symmetry, axial direction and plunge of folds;
(b) the trend, type, and if possible, the throw of faults, recording their

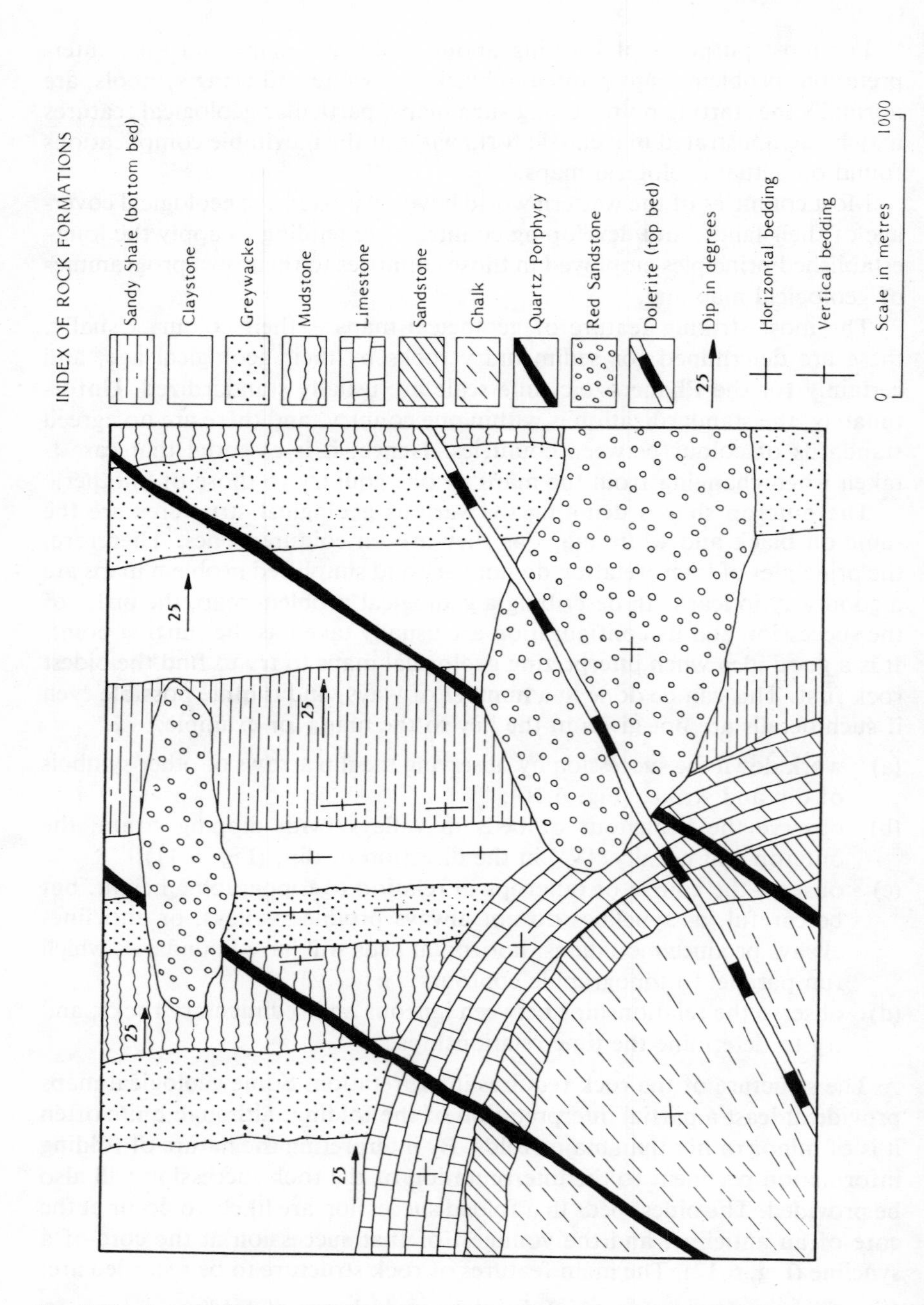

Fig. 6.10 A geological map. (Joint Matriculation Board of the Northern Universities, 1974)

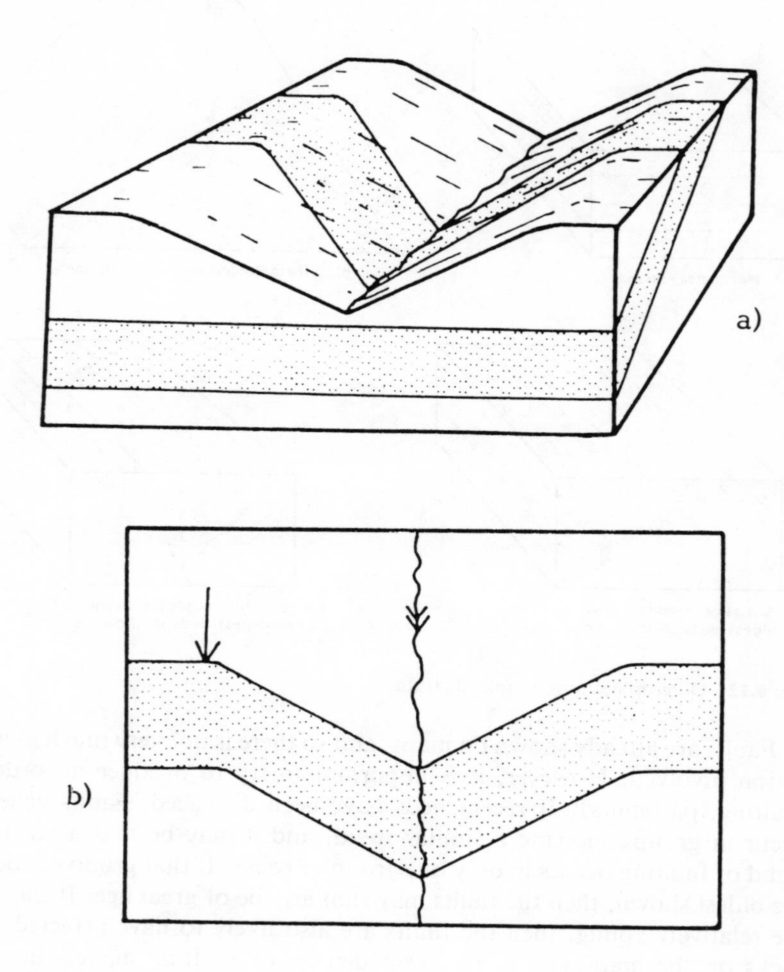

Fig. 6.11 Dip direction and patterns of outcrop in valleys.

relationship to folds, and grouping the faults according to their age and pattern.

(c) the relationships of rock groups, most notably unconformities.

The identification and interpretation of folding may be undertaken by recognizing the characteristic patterns of folded strata as seen at the ground surface. These are:

(a) thin lines of strata trending in the same directions, (Fig. 6.13);
(b) repetition of strata either side of a "core" bed, (Fig. 6.13);
(c) curvature of strata near the "nose" of a fold or made by the erosion of a plunging fold, (Fig. 6.13).

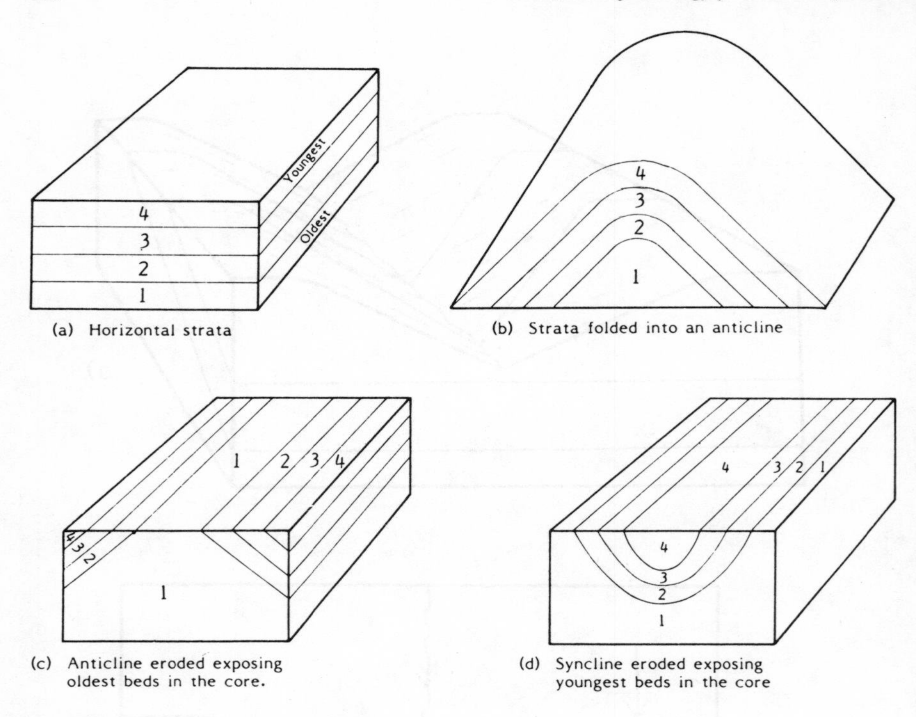

Fig. 6.12 Outcrop patterns of folded strata.

Faults are already shown on maps, and so there is not very much identification involved. However, it is important to try to produce an order of faulting, particularly if recent beds have been displaced. Faults generally occur in groups showing a similar trend, and it may be that a particular trend of faulting occurs in only one group of rocks. If that group of rocks is the oldest shown, then the faults may similarly be of great age. If the rocks are relatively young, then the faults are also likely to have affected older rocks on the map even if the lower density of faulting makes them less apparent. Some faults also appear to end exactly at a rock boundary. In such cases, that may be evidence to suggest that the fault pre-dates the first unfaulted bed. This relationship tends to be most obvious at unconformities and indeed may be used virtually as a definition of one of the many kinds of unconformity (Figs. 6.14a and 6.14b). Since faults displace rocks, they are also likely to displace any other feature which happens to exist at the time, notably other faults. However, faults which are of the same age may cut and displace each other, and you are not likely to find a regular pattern of faults of one particular trend always displacing faults of another trend. Each will cut and displace the other (Fig. 6.15). If there is any age difference between the fault groups, then it may be expected that any displacements will always be of the same type (Fig. 6.16). The unconformities of most

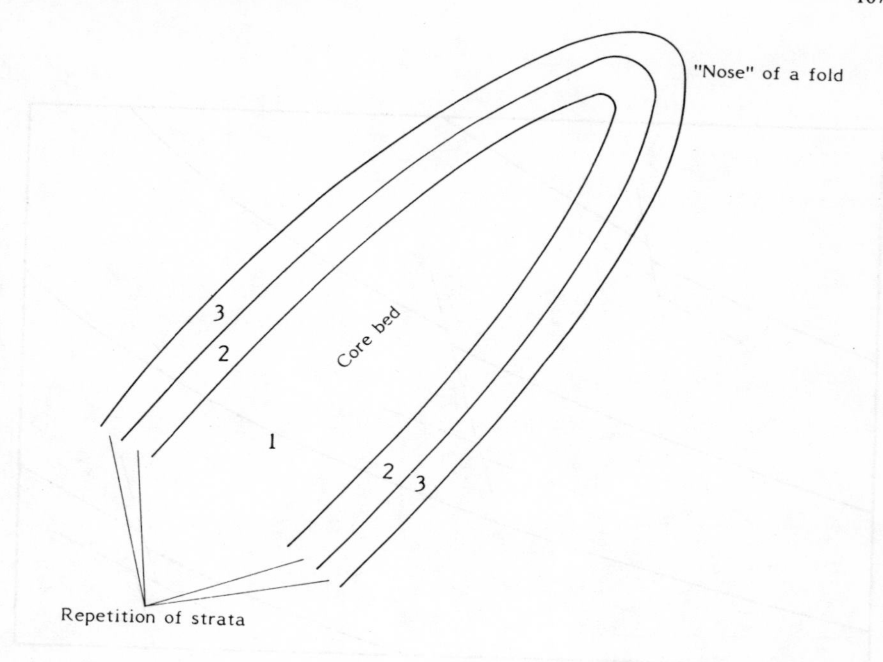

Fig. 6.13 Typical features of folded strata seen on geological maps.

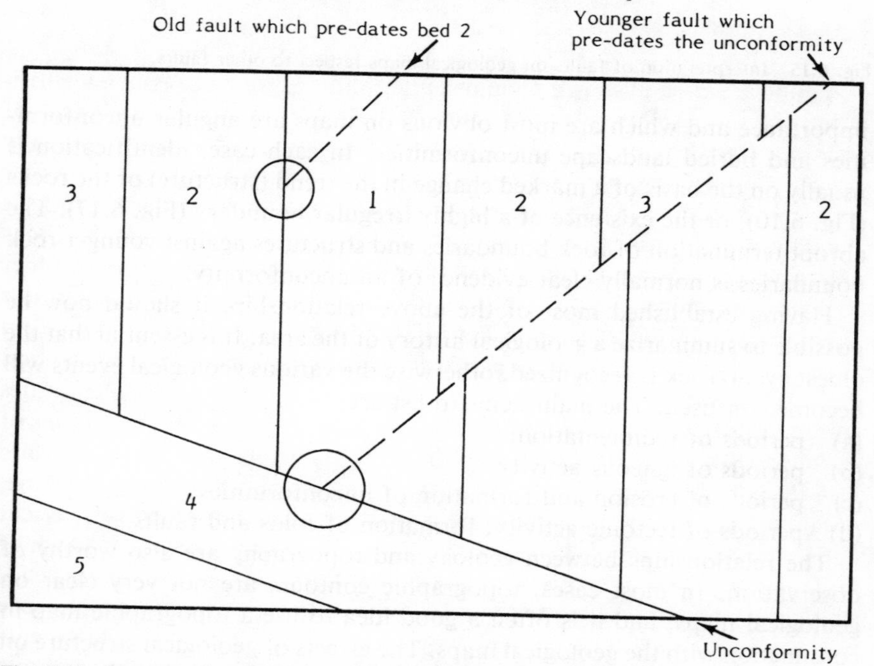

Fig. 6.14 Interpretation of faults on geological maps: age with respect to rock boundaries.

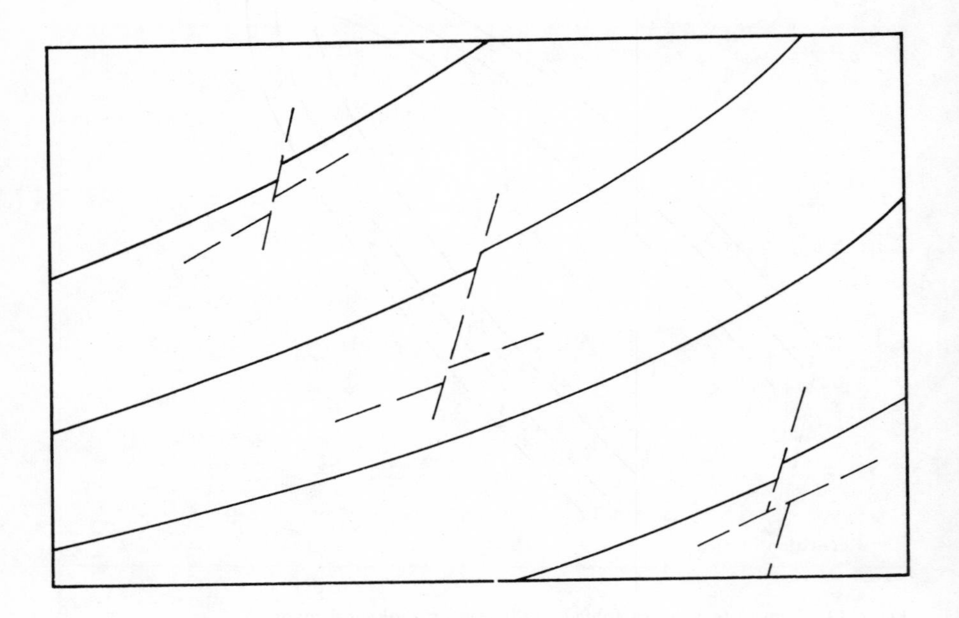

Fig. 6.15 Interpretation of faults on geological maps respect to other faults.

importance and which are most obvious on maps are angular unconformities and buried landscape unconformities. In each case, identification is usually on the basis of a marked change in the trend (structure) of the rocks (Fig. 6.10), or the existence of a highly irregular boundary (Fig. 6.17). The abrupt termination of rock boundaries and structures against younger rock boundaries is normally clear evidence of an unconformity.

Having established most of the above relationship, it should now be possible to summarize a geological history of the area. It is essential that the oldest event/rock is recognized, otherwise the various geological events will become confused. The main items to list are:

(a) periods of sedimentation;
(b) periods of igneous activity;
(c) periods of erosion and formation of unconformities;
(d) periods of tectonic activity; formation of folds and faults.

The relationships between geology and topography are also worthy of observation. In most cases, topographic contours are not very clear on geological maps, and it is often a good idea to use a topographic map in conjunction with the geological maps. The effects of geological structure on scenery may be important, particularly the effects on drainage. Faults may

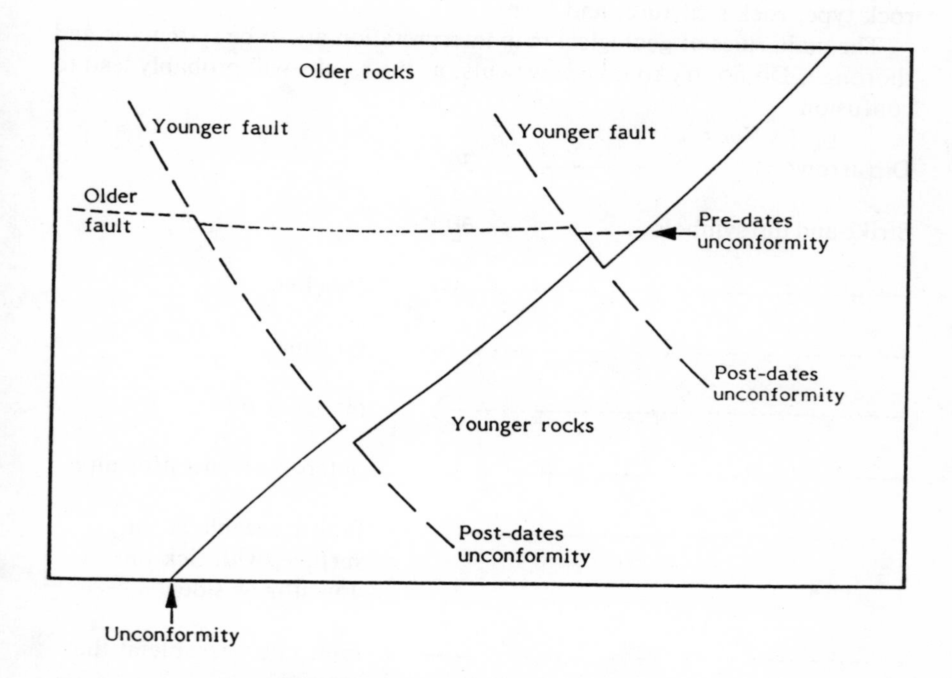

Fig. 6.16 Faults and unconformities on geological maps.

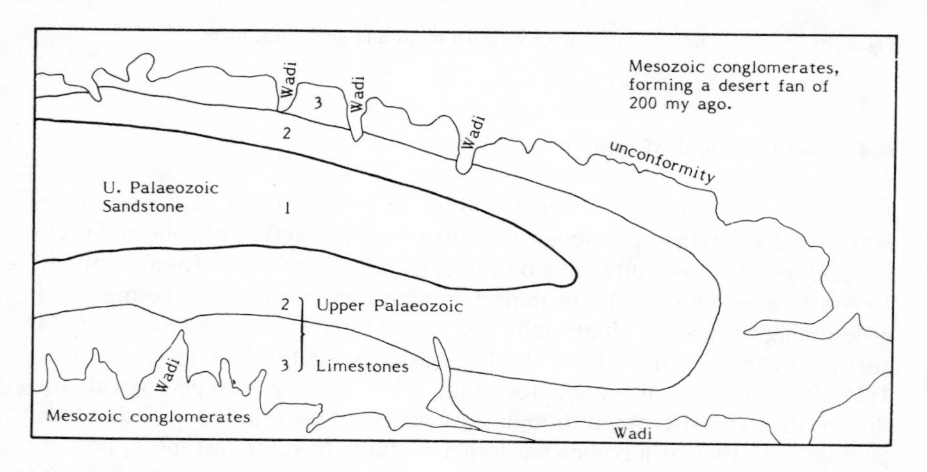

Fig. 6.17 Buried landscape unconformities on geological maps.

have topographic expression and the extent of their erosion may indicate the amount of crushing which took place in the fault zone during its displacement. Coastal areas also tend to show fairly clear relationships between rock type, rock structure, and form.

The main rules of geological map interpretation are to be systematic and thorough. Do not try to take short cuts, as the results will probably lead to confusion.

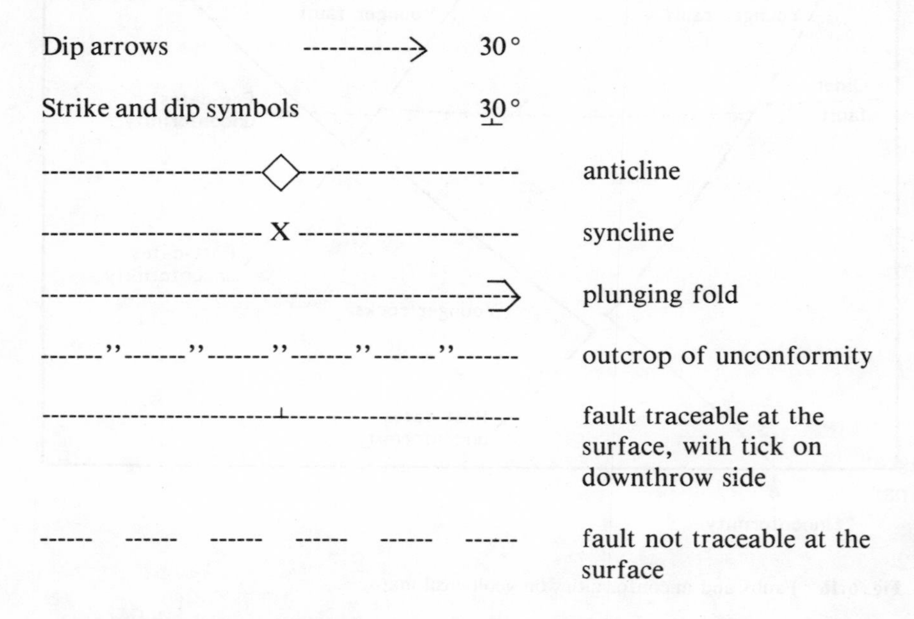

Fig. 6.18 Symbols for use on structural sketch maps and geological maps.

6.4 Geotechnical Mapping

There are a number of shortcomings of conventional geological maps when used as engineering documents. Bracketing together of soils and rocks on geological maps tends to be on the basis of age, origin or fossils content. In most cases, this results in materials with contrasting engineering properties being shown as a single unit. The scale of maps published by geological surveys in many parts of the world is usually 1:50,000 or in some cases, 1:25,000. This sort of scale is too small when dealing with geological conditions on a specific site, although is of course of general use in fitting local geology into that of a somewhat larger picture. In some instances, notably at reservoir sites, such a regional overview is necessary because of the far travel of water leaking from the reservoir. Quantitative data is virtually totally lacking from conventional geological maps. No details of the physical properties of soils and rocks are given. Discontinuity data, apart from a relatively few localized dip and strike measurements, is absent. No impression is given of the thickness of soils or of the weathering mantle. The depth to rock head, the most obvious depth of safe bearing is not indicated. Groundwater conditions can normally only be inferred from a consideration of the rock succession and the topography. Topographic effects are generally difficult to discern because the basic symbols used to interpret topography, contour lines, are frequently obscured beneath the colour on the map. Only in a few cases are foundered or landslipped strata shown. The influence of man is occasionally indicated on some maps, for example the locating of opencast mineral extraction and of mine shafts, although not the pattern of underground mine workings.

A map with a more relevant type of geology is therefore frequently constructed. This is usually called a geotechnical map. The cartographic problems involved in constructing such a map are considerable, and a base map with overlays is a method frequently adopted to overcome them. A variety of different techniques are utilized to construct geotechnical maps.

1 Stratigraphical mapping

This involves the observation, measurement and recording of stratigraphical relationships. Description of strata is in engineering terms with particular attention being paid to lateral and thickness variations.

2 Structural mapping

The orientation, frequency, continuity and morphology of discontinuities are logged. All of these elements have an important effect on engineering behaviour.

3 Geomorphological mapping

The recording of landforms and the processes causing them represents a record of recent geological history in an area. It is likely to be of particular significance to engineers in indicating the nature of deposits of the recent geological past (a few million years), an aspect of geological maps which is normally poorly covered. Furthermore, the logging of current processes may indicate how disturbance to the landscape is likely to be manifested, and the nature of the response.

4 Hydrogeological mapping

This involves the location of groundwater basins, artesian overflow, groundwater movement, water tables, water chemistry. Rocks may also be classified as aquifers, that is significantly water bearing and permeable, and aquicludes, where permeabilities are generally low.

5 Geomechanical mapping

Soil and rock mechanics data can be mapped if it exists in sufficient quantity. This is possible within the confines of most engineering sites, and has in fact been done regionally.

6 Mapping the effects of man

Mining and quarrying activity, infilled dumps, made ground, subsidence,

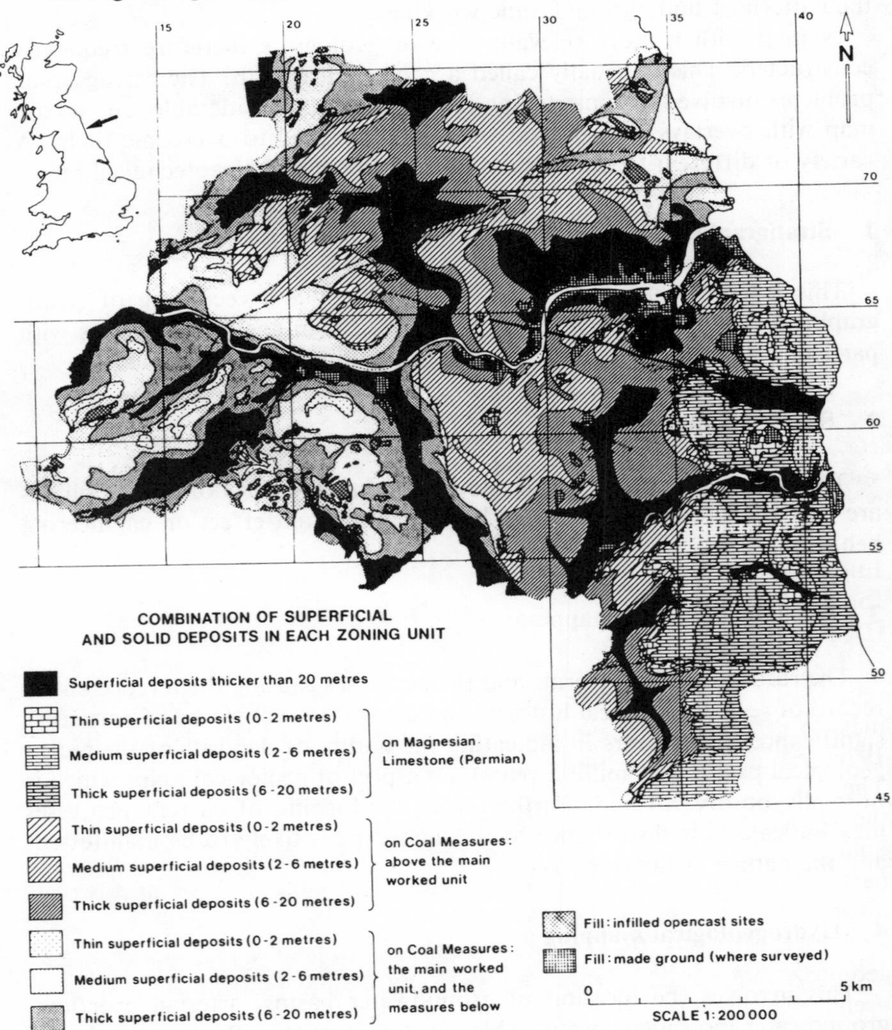

Fig. 6.19 Zoning of soils on engineering geological maps. (Dearman, W.R. *et al.*, 1979; Copyright © 1979 by *International Association of Engineering Geology*)

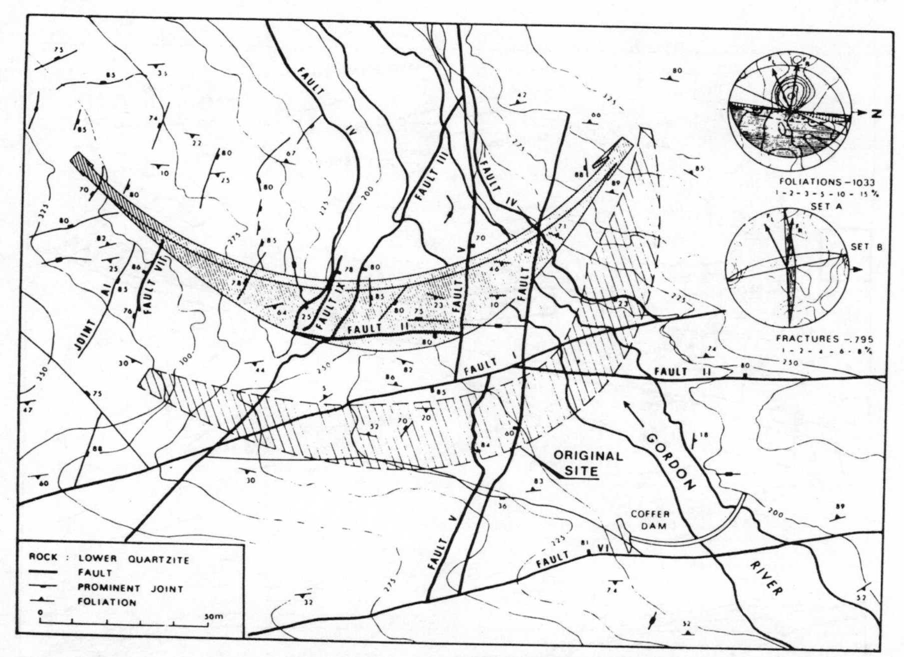

Fig. 6.20 Structural geology map for an engineering project. (Andric, M., Roberts, G.T. and Tarvydas, R.K., 1976; Copyright © 1976 by *Geological Society of London*)

etc. all have potential implications for engineering activity by causing sudden changes in ground conditions.

An example of stratigraphical mapping is shown in Fig. 6.19. As is evident from the legend, man's interference in this part of North-East England is important to the civil engineer, and the effects of mining have formed one of the criteria upon which the mapping of soils has been based. (Dearman *et al.*, 1979)

The two maps of geological structure (Figs. 6.20 and 6.21) are both related to dam projects and indicate the different scales at which mapping may be carried out. In Fig. 6.20, for a dam in Tasmania, (Andric *et al.*, 1976) the faults were a particularly important aspect of the rock structure. The original site proposed for the dam was moved so that it did not exist in the area of faulting, which was found to be related to markedly poorer rock quality at junctions with other faults. In Fig. 6.21, the details of structure in part of a dam spillway are shown, (Anon, 1972) and a variety of different discontinuities of significance are plotted.

Since the early 1970s, more attention has been paid to the significance of geomorphology in civil engineering. Many geomorphological maps have been constructed as part of the site investigation for a variety of projects. It has been found to be particularly useful in developing countries. Two examples of geomorphological mapping are presented. The first (Fig. 6.22a)

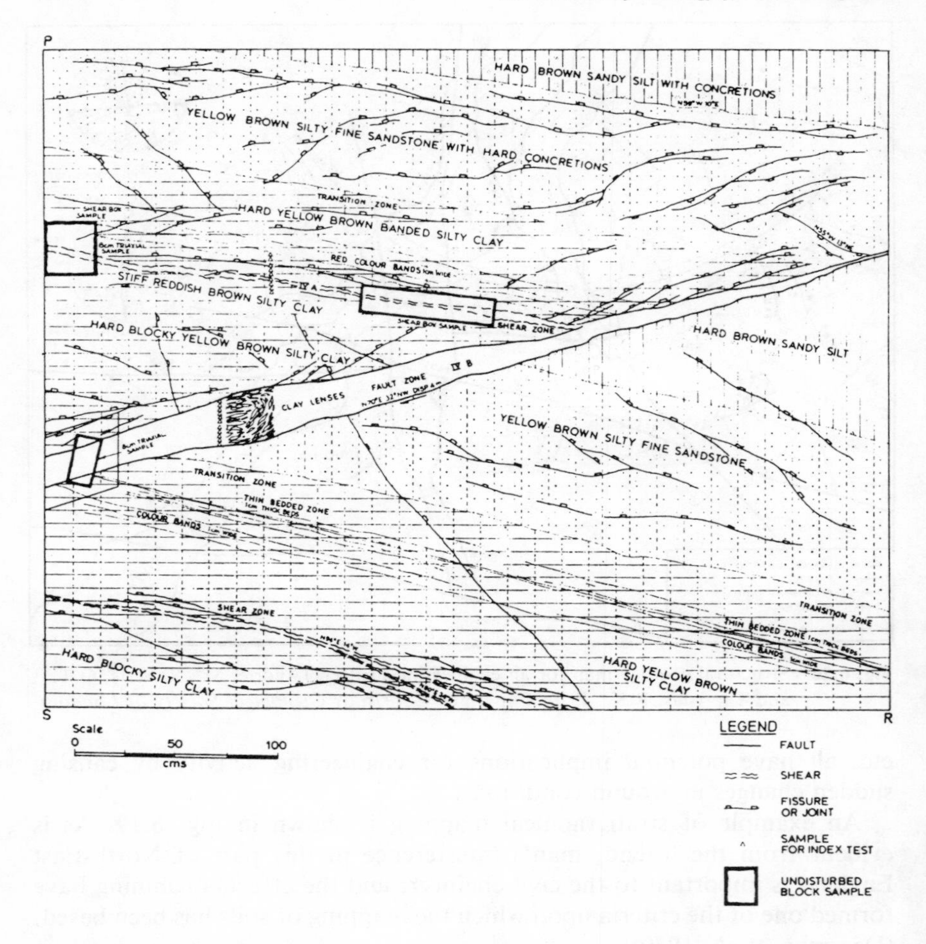

Fig. 6.21 Structural geology map for part of an engineering project. (Anon, 1972; Copyright © 1972 by *Geological Society of London*)

is for a single large rotational failure in rock in the south coast of England (Pitts, 1979). This wealth of detail can be simplified and presented as a slope categories map, (Fig. 6.22b). In Fig. 6.23a a geomorphological map is shown which was constructed to aid in the design of coast protection works (Pitts, 1983a). Both forms and processes are marked on this map. Slope angle details are not shown, and a series of profiles, with relevant related information (Fig. 6.23b) is used in conjunction with the map.

Many aspects of hydrogeology may be mapped. The form of the water table is of considerable significance and may be mapped in the same way as any other surface, namely using contours, (Fig. 6.24).

Geomechanical mapping is normally undertaken for individual formations to indicate changes in soil or rock mechanics properties with depth and

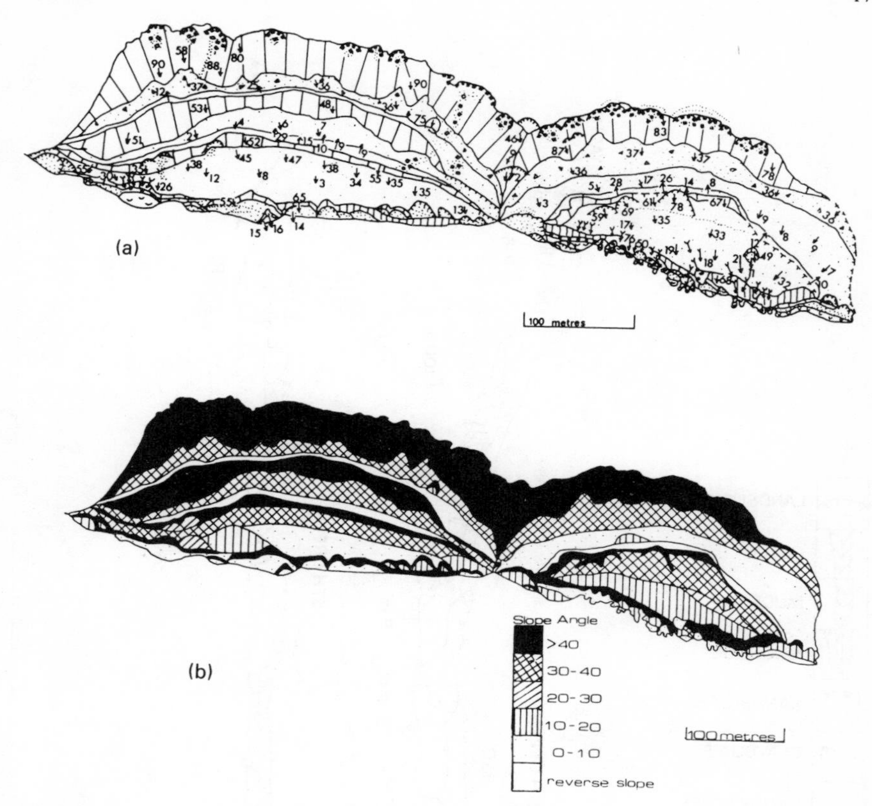

Fig. 6.22 (a) Geomorphological maps and,
(b) Slope categories maps for engineering usage.
(Pitts, J., 1979; Copyright © 1979 by *Geological Society of London*)

laterally. A large-scale project on the London Clay was carried out (Burnett and Fookes, 1974) and two maps from a series of geomechanical maps are shown, (Figs. 6.25 and 6.26). The relationship between trends in such index properties as liquid limit and the physical composition of the soil, in this case, clay content, show clear correlation.

The significance of the site investigation stage of a project is clearly shown in the maps illustrating the effects of man, in this case in an urban environment, (Figs. 6.27a and b). The proposed layout of a housing development in Edinburgh, Scotland (Price *et al.*, 1969), is shown in Fig. 6.27a. This was before the presence of an infilled sand working, infilled limestone quarry and worked coal seams were taken into account. By carefully mapping these features, the optimum type of structure and their location could be presented (Fig. 6.27b).

A large amount of information relevant to a civil engineering project tends to come to light as a result of a site investigation. A map presents in a

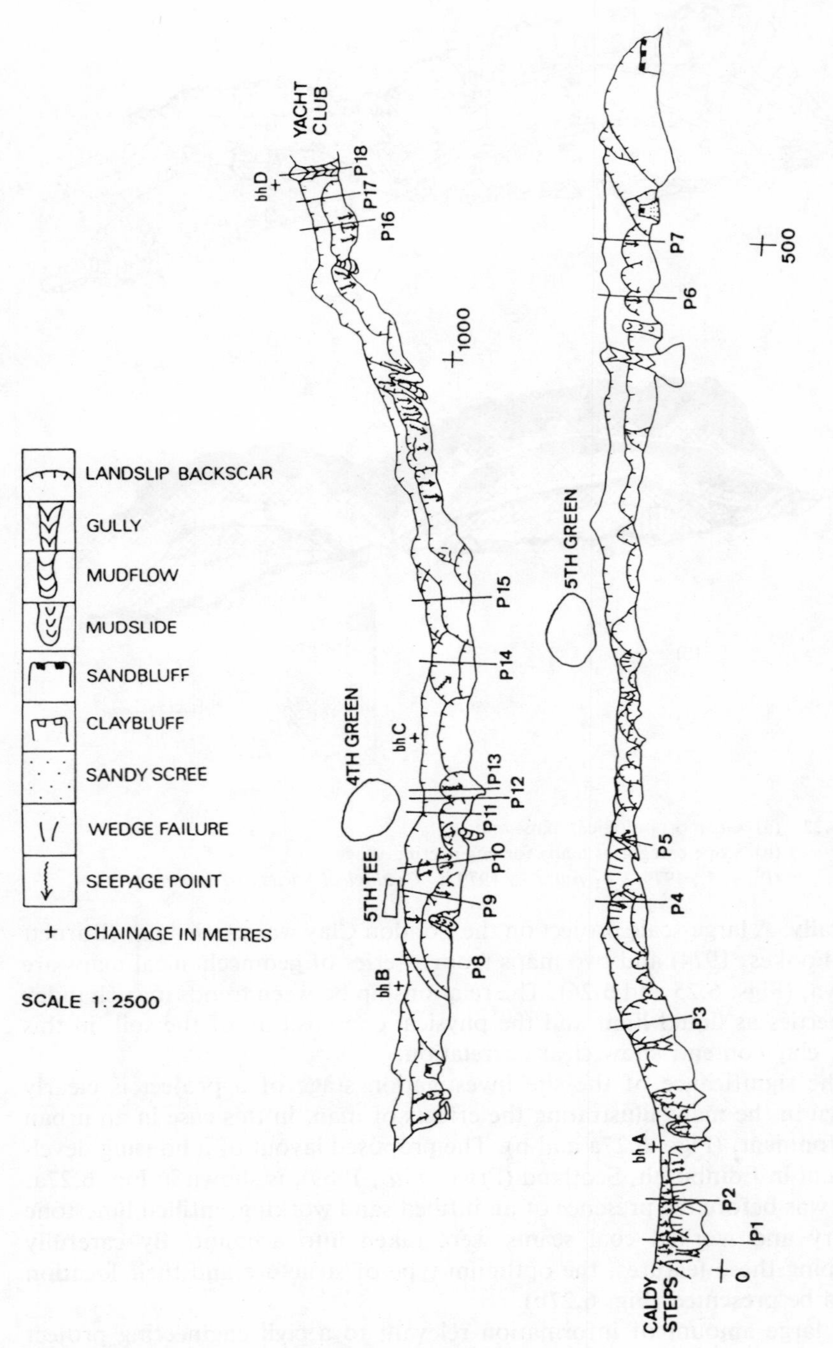

Fig. 6.23 (a) Geomorphological maps for engineering use.

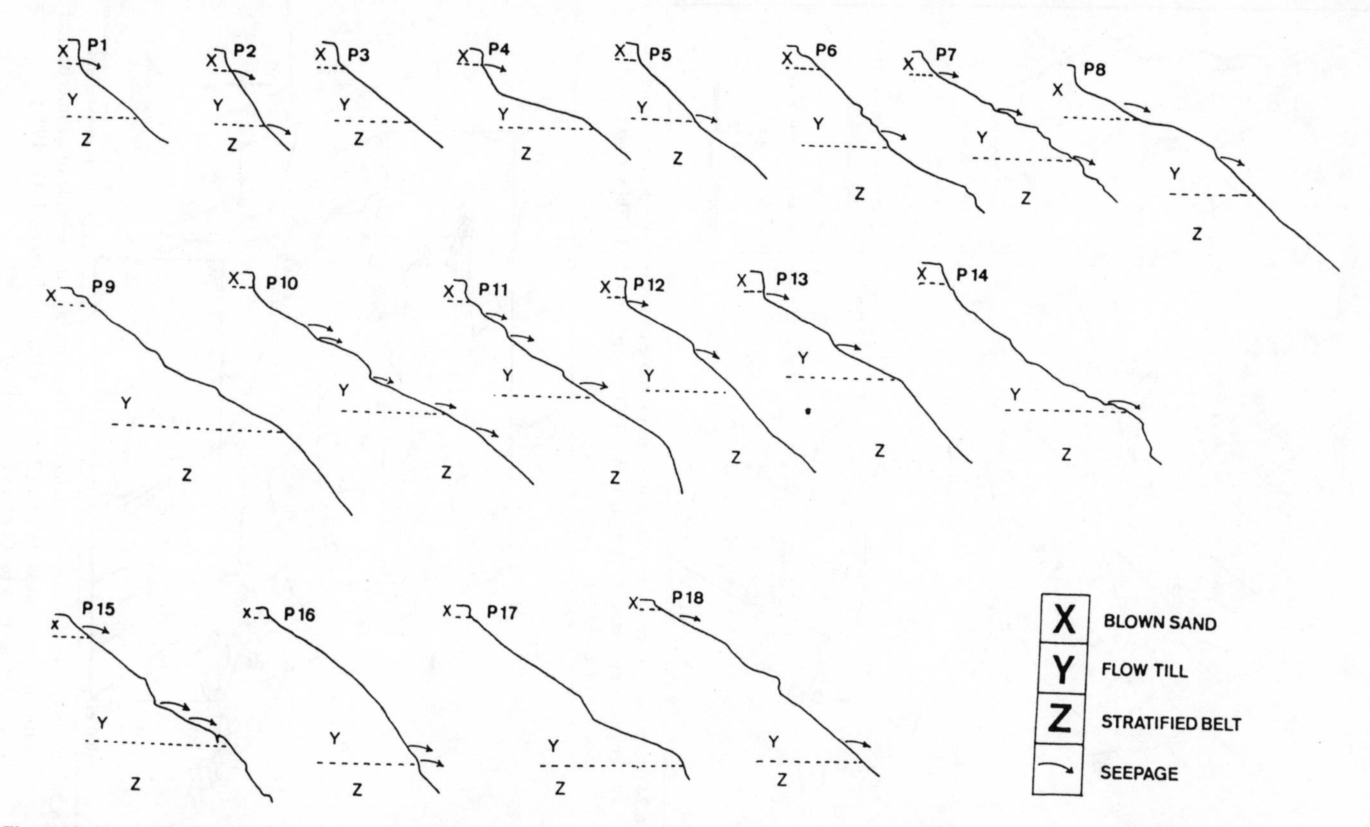

Fig. 6.23 (b) Topographic profiles for use with engineering geomorphological maps. (Pitts, J., 1983a; Copyright © 1983a by *Geological Society of London*)

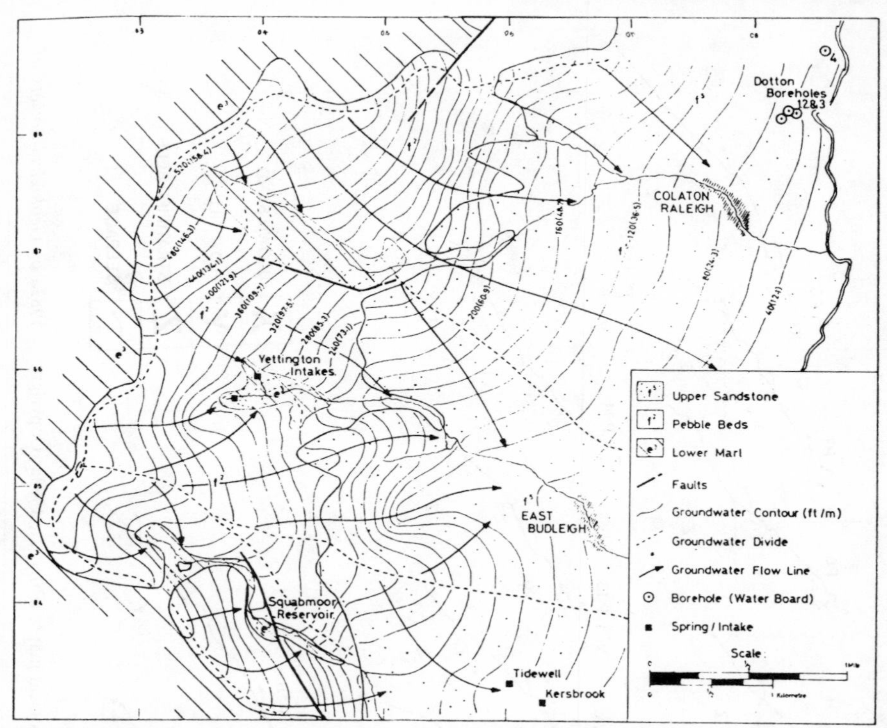

Fig. 6.24 Contours of groundwater tables: an example of hydrogeological mapping for engineering usage. (Sherrell, F.W., 1970; Copyright © 1970 by *Geological Society of London*)

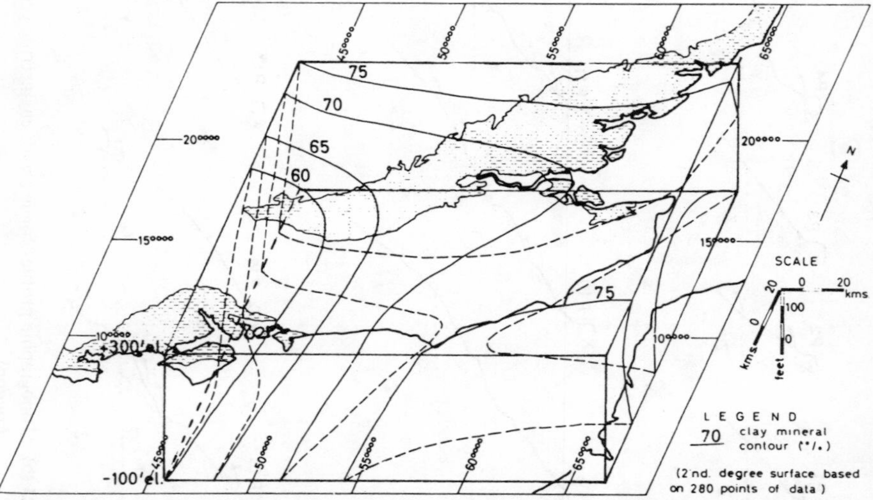

Fig. 6.25 Contours of clay mineral content of London clay: an example of geomechanical mapping for engineering use. (Burnett, A.B., and Fookes, P.G., 1974; Copyright © 1974 by *Geological Society of London*)

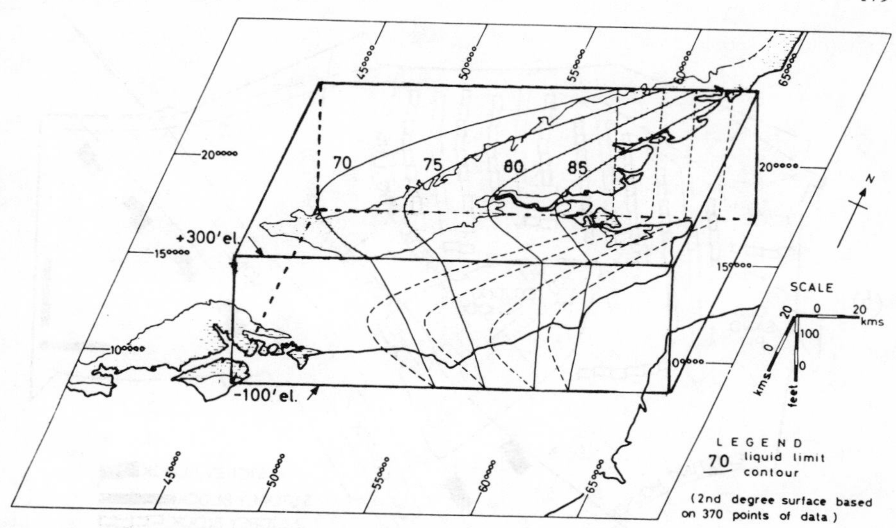

Fig. 6.26 Contours of variation in liquid limit of London clay: an example of geomechanical mapping for engineering use. (Burnett, A.B. and Fookes, P.G., 1974; Copyright © 1974 by *Geological Society of London*)

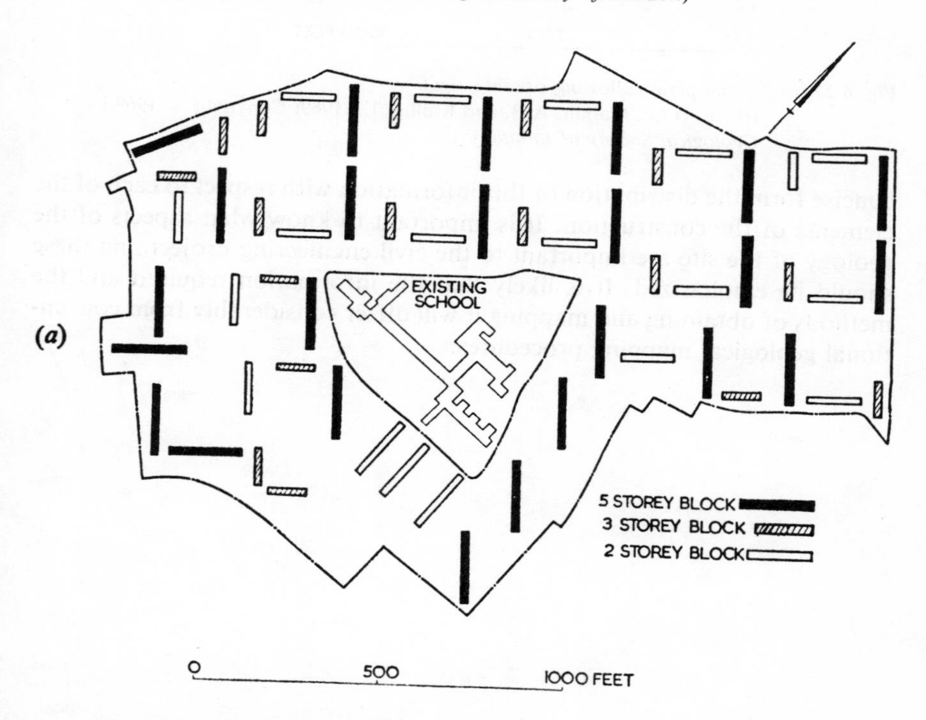

Fig. 6.27 Mapping the effects of man on a geological environment in an urban setting: (a) proposed plans for a development.

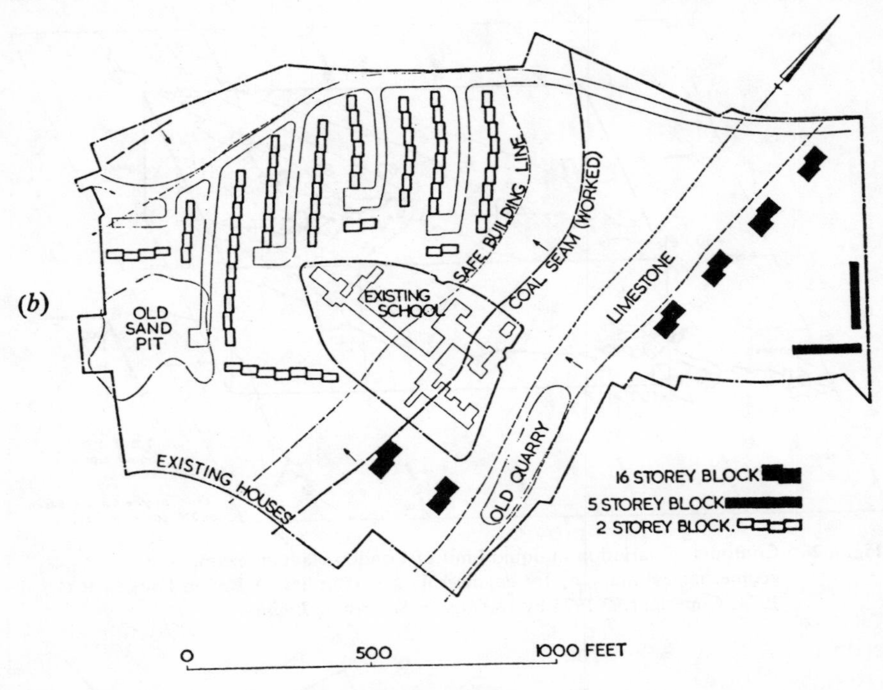

Fig. 6.27 (b) actual plans following site investigation. (Price, D.G., Malkin, A.B. and Knill, J.L., 1969; Copyright © 1969 by *Geological Society of London*)

concise form the distribution of this information with respect to each of the elements of the construction. It is important to know what aspects of the geology of the site are important to the civil engineering project and these should be emphasized. It is likely that the information required and the methods of obtaining and mapping it will differ considerably from conventional geological mapping procedures.

CHAPTER 7: LOGGING ROCKS FOR ENGINEERING PURPOSES

7.1 Logging of Rock Cores

1 General requirements of a borehole log

A borehole log should provide an accurate and comprehensive record of the geological conditions encountered together with any other relevant information obtained during drilling. Although many individuals still use their own system of core logging, the introduction of recommended procedures by the Engineering Group of the Geological Society of London (Anon, 1970) has generally increased the uniformity of logging methods during the past decade. This has enhanced the 'readability' of logs and reduced the problems of interpretation.

2 Information to be recorded

The information to be recorded can be summarized under the following headings:

(a) Basic information including project name, location, timing, Employer, Engineer and Contractor;
(b) Drilling method and progress;
(c) Description of type and condition of material encountered;
(d) Key and miscellaneous comments.

The basic information provided must be adequate to locate the borehole in time and space, to define the authorities responsible for all aspects of the boring and to assist data storage and retrieval. The project name and locality together with the name of the drilling contractor or sub-contractor should be provided.

The machine, core barrel and bit should all be described. Detailed records of core sizes and changes, use of casing, standing water levels, losses of circulating fluid and penetration rate can also provide useful supplementary information on the ground conditions. The main part of the log consists of the description of the geological features of the core relative to a depth scale. The condition of the sample recovered should be defined in terms of both the proportion actually recovered and the state of fragmentation. The descriptive geology should include definition of the rock type, its alteration state and relative strength, together with information on the natural discontinuities and rock structures.

Once the cores are disposed of, the borehole log remains the sole tangible product of an expensive operation; this factor should not be forgotten. The sole final record of drilling should be a comprehensive log. The drillers' daily record sheet is an important source of technical and also contractual information, and use should be made of this.

3 Information to be recorded on the borehole log

The essential information which needs to be recorded on the log is as follows: Borehole number; this should be used only once on a site and kept as simple as possible without extraneous ciphers.

Location:

(a) Site, including project name, town, country or state name where necessary.
(b) Grid Reference of the borehole should always be stated to at least 10 m accuracy.
(c) Elevation relative to O.D. for the ground or sea bed level or underground datum at the borehole site to an accuracy of 0.1 m.
(d) Orientation of the borehole given as an angle to the horizontal (+ve upwards, −ve downwards) and azimuth (0° to 360° clockwise relative to Grid North).

Drilling technique: The following should be stated . . .

(a) The method of penetration and flush system.
(b) The make of machine with the model number and feed type if other than hydraulic.
(c) The type of core barrel and bit.

Contract details: The following should be noted . . .

(a) Name of site investigation contractor
(b) Name of client or authority
(c) Job Reference number
(d) Name or cypher of logger

Miscellaneous: There should be an opportunity for relevant miscellaneous information to be included in the log.

Apart from the data listed above, the following technical information should be recorded by the driller on his daily record sheet:

(a) Depth of hole at start and end of working day or shift as relevant.
(b) Depth of start and finish of each core run.
(c) Depth and size of casing at start and end of each core run.
(d) Core diameter and changes in core size.

(e) State of bit.
(f) Time to drill each core run.
(g) Character and proportion of flush return.
(h) Standing water level at start and end of each working period.
(i) Simplified description of strata.
(j) Total core recovery with information as to possible location of core losses.
(k) Sample locations.
(l) Details of delays and breakdown.
(m) Details of in situ tests and instrumentation installed.
(n) Backfilling and grouting.

This list excludes items which may be required for contractual record purposes.

Particular mention should be made of the value in the detailed study of flush returns and standing water levels during drilling. Loss of water during drilling is a useful measure of permeable conditions as are variations in the "make" of water during air-flush drilling. Standing-water levels should be carefully related to the ground conditions and location of the casing; it is important to ensure that such levels have become adequately stabilized.

4 State of recovery of core

The state of the rock cores recovered is a valuable indication of the in situ condition of a rock mass and its probable engineering behaviour. This must be assessed against the drilling methods used and, on individual projects, it is desirable that standardization of drilling technique should be adopted to permit comparison. In any core recovered, fractures may be of natural or artificial origin. Even in the soundest rock, some artificial fragmentation (particularly at the ends of the core runs) is to be expected and should be recognized.

During the drilling process, the bit cuttings are removed in the flush system. The sample which passes up into the core barrel may be divided into five parts:

(a) solid core greater than 0.1 m in length (see discussion of RQD below);
(b) solid core less than 0.1 m in length;
(c) fragmental material not recovered as core;
(d) soft or friable material eroded and removed by the flushing system (sometimes resulting in a reduction in diameter rather than length of the core) and
(e) additional material which may have been lost from the previous core run. This may be the core stump left when the barrel was pulled, material dropped from the core barrel during its withdrawal from the hole or cuttings which have settled when circulation stopped.

The material which is placed in the core box consists of items (a), (b), (c) and (e) above and (omitting (e) from the subsequent discussion) is strictly defined as the total core recovery. If no material falls into class (d), then the *total core recovery* is 100% in that there is no loss of sample. The material which is recovered as solid core pieces at full diameter ((a) and (b) above) is strictly defined as the *solid core recovery*. It must be stressed that total and solid core recoveries are only equivalent when no fragmental material is recovered which arises either when the rock is solid or loss of sample is represented by material wholly carried away in the flushing system. It should be noted that recoveries are, for convenience, measured in terms of length and expressed relative to core run length.

Various criteria may be used for quantitative description of the fracture state of the cores; these are the solid core recovery, fracture log, RQD and Stability Index. The simplest measure is the solid core recovery, particularly when contrasted to the total core recovery. A fracture log is a count of the number of natural fractures present over an arbitrary length; it is conventional to omit obvious artificial fractures and discontinuities resulting from deterioration of the core.

The solid core recovered can be separated into fractions greater and less than 100 mm in length. Some support is provided for a correlation between the percentage solid core recovered greater than 100 mm in length (the Rock Quality Designation = RQD) and various physical parameters of the rock mass.

7.2 Descriptive Geology

The following factors have to be incorporated in a log for adequate engineering geology description:

(a) systematic description
(b) alteration/weathering state
(c) structure and discontinuities
(d) assessment of rock material strength
(e) other features, including stratigraphy

The following standard sequence of systematic description is proposed:

(i) weathered state
(ii) structure
(iii) colour
(iv) grain size (a) subordinate particle size
(b) texture
(c) alteration state as relevant
(v) rock material strength
(vi) (a) mineral type as relevant — ROCK NAME

It is considered that the qualifications are more important in core descriptions than the actual rock name and, for this reason, the name was placed last. Such a system is appropriate to an engineering description whereby classification by mechanical properties is more appropriate than classification by mineralogy and texture. The following example is provided for illustrative purposes:

Completely weathered	thinly flow-banded	mid-grey	very coarse
(i)	(ii)	(iii)	(iv)

porphyritic	kaolinized	weak	tourmaline	GRANITE
iv.a	iv.c	v	vi.a	vi

The weathering grade should be assessed according to Fig. 4.2. A scale of strength, based on uniaxial compressive tests, is shown in Table 7.3.

Any rock with a strength significantly less than 2.5 MN/m^2 should be described with reference to soil mechanics practice.

It should be noted that field assessments of rock strength can be aided by simple tests, such as the use of a hammer or penknife, and readily supplemented by the Schmidt hammer.

The systematic description of the rock cores may need to be supported by ancillary geological information aiding correlation between boreholes. This might include, for example, dip, stratigraphical location, sedimentary structure, identification of fossil bands and coal seams or qualification relating to genesis. The location of each natural fracture should be shown in the symbolic log, along with the ancillary information. An example of a completed borehole log is shown in Fig. 7.1.

7.3 The Description of Rock Masses for Engineering Purposes

Standardization of technique in describing rock masses is desirable in aiding communication between various members of the civil engineering industry. Consideration of a few basic parameters to produce a rock mass "rating" representing a quick guide to engineering behaviour may assist in this development. Parameters equivalent to soil mechanics index properties, some descriptive, others of a (quasi-) quantitative nature may help to achieve this aim. Indices describing intact rock and discontinuities are required. A number of methods of rock mass logging exist, but the method recommended by the Engineering Group of the Geological Society of London (Anon, 1977) will be covered here.

1 Rock material description

Rock material may be described in two ways:

(a) petrographically, emphasizing mineralogy, grain interaction, geological history;

or (b) in terms of engineering properties.

Method (b) is recommended, although some basic petrography is essential.

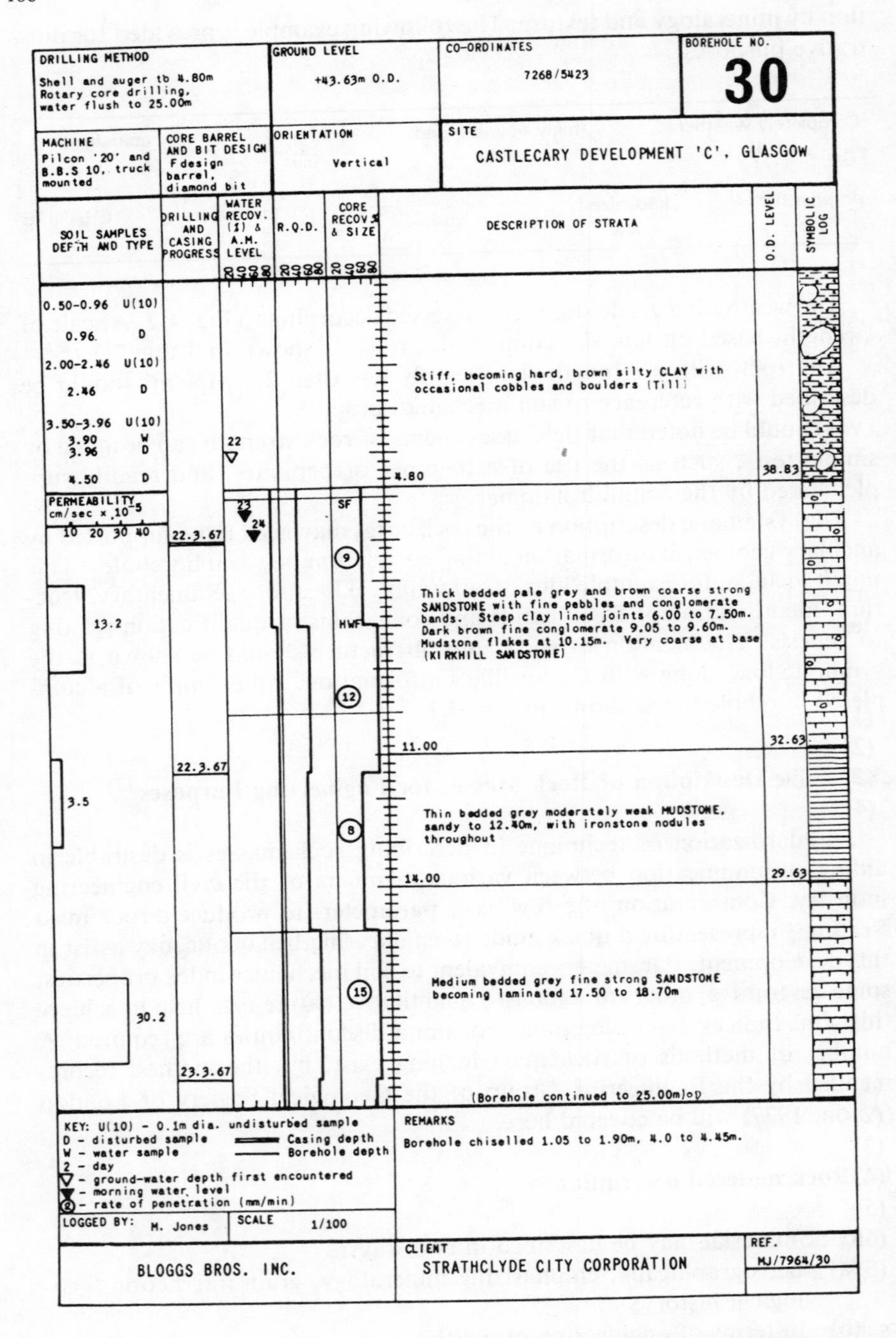

Fig. 7.1 Example of a completed rock core log. (Anon, 1970; Copyright © 1970 by *Geological Society of London*)

2 Rock material indices

The following indices provide a full description of the rock material:

Group 1		Group 2	
Rock Type Colour Grain size Texture and Fabric Alteration Strength	Descriptive Indices	Hardness Durability Porosity Density Strength Seismic velocity	Indices which can be determined by classification tests requiring little or no sample preparation
Group 3			
Young's Modulus of Elasticity Poisson's Ratio Primary Permeability		Indices for design normally only determined by complex testing, or requiring extensive sample preparation or both	

Properties in Group 1 are descriptive; those in Group 2 may be determined by simple classification tests; and those in Group 3 are determined by testing generally in a laboratory.

Description of rock type should be according to the following standard sequence:

(1) weathered state
(2) structure
(3) colour
(4) grain size
  (a) subordinate particle size
  (b) texture
  (c) alteration state
  (d) cementation state, if relevant
(5) rock material strength
(6a) mineral type, as relevant
(6) ROCK NAME

An example of this may be

(1) Fresh
(2) foliated
(3) dark grey
(4) coarse
(5) very strong
(6a) hornblende
(6) GNEISS

Determination of colour may be by colour chart, or by the simplified scheme shown in Table 7.1. A colour in Column 3 may be supplemented, if necessary, by a term from Column 2, or Column 1, or both.

Table 7.1 Rock colour.

1	2	3
light	pinkish	pink
dark	reddish	red
	yellowish	yellow
	brownish	brown
	olive	olive
	greenish	green
	bluish	blue
		white
	greyish	grey
		black

Grain size determination is by the scheme in widespread use shown in Table 7.2.

The strength of the rock is only important in its influence on the shear strength of discontinuities or in unfractured rock. The strength of the rock mass is largely governed by discontinuities. Strength determinations may be carried out with a point load strength tester or Schmidt impact hammer and may be expressed as shown in Table 7.3, which also gives a guide to field assessment of strength.

3 Rock mass indices

Following the description of the rock material, it is necessary to recognize and describe the features dividing the mass. The following are guides to the description of discontinuities.

Group 1	
Discontinuities	
type	
number of orientations	
location and orientation	
frequency of spacing between discontinuities	Descriptive
aperture or separation of discontinuity surface	Indices
persistence and extent	
infilling	
nature of surfaces	
additional information	
weathered and altered state	
Group 2	
Permeability (primary)	Indices which can be determined by relatively simple
Seismic velocity	classification tests
Shear strength	

Group 3	
Modulus of deformability Permeability (secondary) Seismic velocity Shear strength	Indices for design normally only determinable by complex testing

Table 7.2 Grain size.

Term	Particle size	Retained on B.S. Sieve No. (approx. equivalent)	Equivalent Soil Grade
Very coarse-grained	> 60 mm	2 in	Boulders and Cobbles
Coarse-grained	2–60 mm	8	Gravel
Medium-grained	60 microns–2 mm	200	Sand
Fine-grained	2–60 microns		Silt
Very fine-grained	< 2 microns		Clay

Table 7.3 Rock material strength.

Term	Unconfined compressive strength MN/m² (MPa)	Field estimation of hardness
Very strong	> 100	Very hard rock — more than one blow of geological hammer required to break specimen
Strong	50–100	Hard rock — hand held specimen can be broken with single blow of gelogical hammer
Moderately strong	12.5–50	Soft rock — 5 mm indentations with sharp end of pick
Moderately weak	5.0–12.5	Too hard to cut by hand into a triaxial specimen
Weak	1.25–5.0	Very soft rock — material crumbles under firm blows with the sharp end of a geological pick
Very weak rock or hard soil	0.60–1.25	Brittle or tough, may be broken in the hand with difficulty
Very stiff	0.30–0.60	Soil can be indented by the finger nail
Stiff	0.15–0.30	Soil cannot be moulded in fingers
Firm	0.08–0.15	Soil can be moulded only by strong pressure of fingers
Soft	0.04–0.08	Soil easily moulded with fingers
Very soft	< 0.04	Soil exudes between fingers when squeezed in the hand

7.4 Discontinuities

A discontinuity is considered to be a plane of weakness within the rock across which the rock material is structurally discontinuous and has zero or low tensile strength, or a tensile strength lower than the rock material under the stress levels generally applicable to engineering. Thus, a discontinuity is not necessarily a plane of separation.

Discontinuities may occur in sets (e.g. joints, cleavages, bedding planes), or be unique (e.g. faults). Differentiation of genesis is important in establishing engineering properties, e.g. origin in extension or shear.

The properties of a rock mass (e.g. strength, deformability, permeability) will be determined by number, number of sets and spacing of discontinuities these governing failure and deformation without involving fracture of rock material.

Orientation of discontinuities is determined in terms of dip and dip direction (measured to the nearest degree) by a compass and clinometer. Spacing is best measured as mean fracture spacing (per metre) along a traverse line, preferably in more than one direction, and expressed as shown in Table 7.4.

The degree of openness or separation of discontinuities is very important and should be measured and described as in Table 7.5. Infilling along discontinuities (gouge, mineralization etc.), may influence shearing resistance, and should be measured and described. The unconfined compressive strength of the infill should be determined by pocket penetrometer (soils) or Schmidt Hammer/Point load tester (rocks). The nature of the walls of the discontinuities should also be described.

Persistence of discontinuities is extremely difficult to describe and quantify. In general, maximum trace length should be measured, with comment as to whether termination is in solid rock or against another discontinuity.

The nature of discontinuity surfaces, viz. wavy or rough should be described and measured by rule, where possible. Measurement of roughness may best be done by referring to the categories described in Table 7.6.

Additional information, where applicable, should be recorded, particu-

Table 7.4 Discontinuity spacing.

Term	Spacing
Extremely wide	>2 m
Very wide	600 mm-2 m
Wide	200-600 mm
Moderately wide	60-200 mm
Moderately narrow	20-60 mm
Narrow	6-20 mm
Very narrow	<6 mm

larly with reference to groundwater seepage, presence of swelling materials, indications of instability.

Table 7.5 Aperture of discontinuity surfaces.

	Aperture (discontinuities) Thickness (veins, faults)
Wide	> 200 mm
Moderately wide	60–200 mm
Moderately narrow	20–60 mm
Narrow	6–20 mm
Very narrow	2–6 mm
Extremely narrow	> 0–2 mm
Tight	zero

Table 7.6 Roughness categories.

Category	Degree of roughness
1	Polished
2	Slickensided
3	Smooth
4	Rough
5	Defined ridges
6	Small steps
7	Very rough

1 Weathering and altered state

Weathering of the rock and the discontinuities should be described. Weathering grades are shown in Table 4.2, but may be modified where appropriate.

2 Discontinuity spacing in three dimensions

The following terms define shape:
blocky — approximately equidimensional
tabular — one dimension considerably shorter than the other two.
columnar — one dimension considerably larger than the other two.

In tabular and columnar types, specify the orientation of the long or short dimension, and possibly also the ratio of orthogonal dimensions, e.g. 1 vertical: 2 north: 6 east.

Block size may be described as shown in Table 7.7. Charts for logging rock masses in the field are shown in Figs. 7.2 and 7.3.

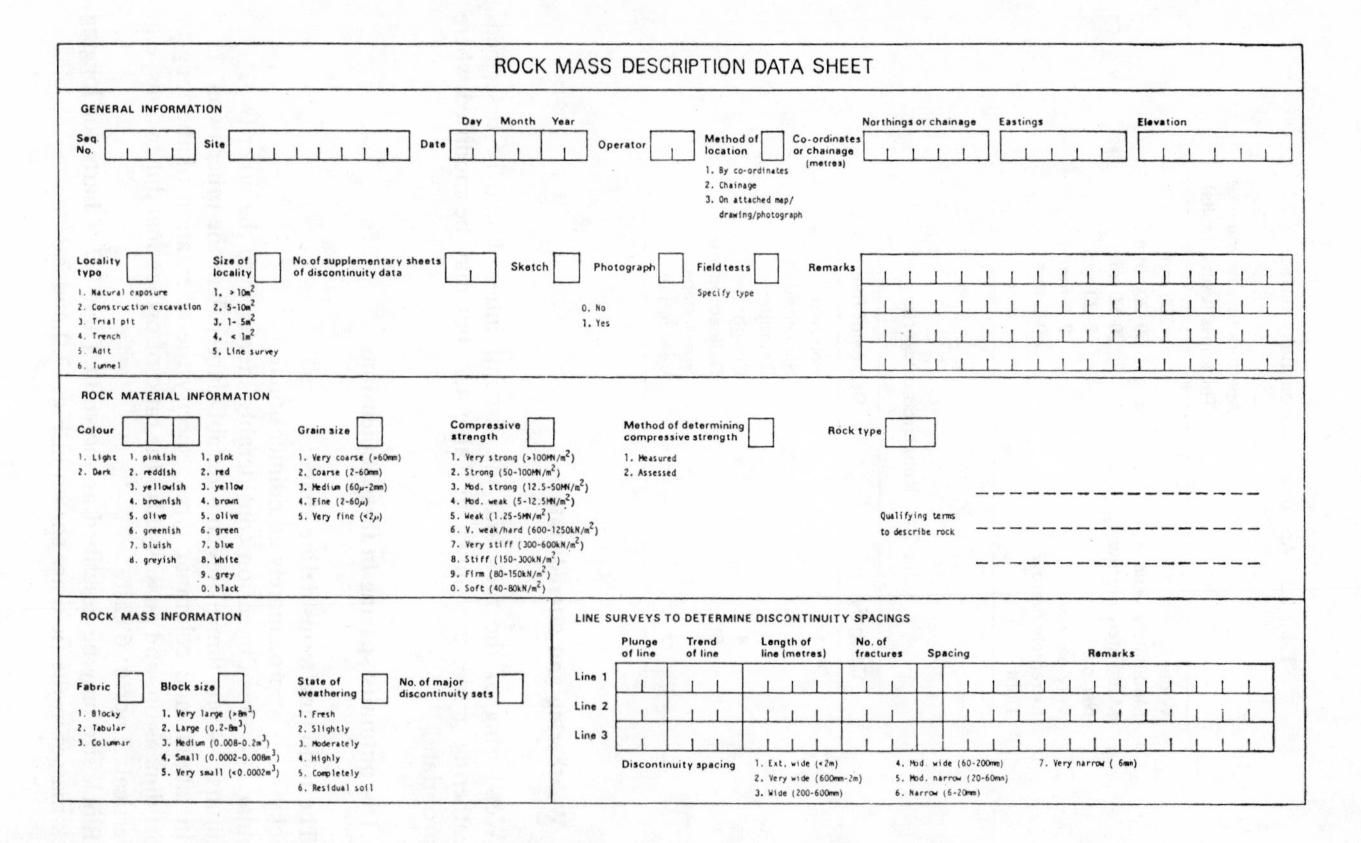

ROCK MASS DESCRIPTION DATA SHEET

GENERAL INFORMATION

Seq. No. | Site | Date (Day, Month, Year) | Operator | Method of location | Co-ordinates or chainage (metres) | Northings or chainage | Eastings | Elevation

Method of location:
1. By co-ordinates
2. Chainage
3. On attached map/drawing/photograph

Locality typo
1. Natural exposure
2. Construction excavation
3. Trial pit
4. Trench
5. Adit
6. Tunnel

Size of locality
1. > 10m²
2. 5-10m²
3. 1- 5m²
4. < 1m²
5. Line survey

No. of supplementary sheets of discontinuity data | Sketch | Photograph | Field tests | Remarks

Sketch / Photograph:
0. No
1. Yes

Field tests: Specify type

ROCK MATERIAL INFORMATION

Colour
1. Light
2. Dark

1. pinkish
2. reddish
3. yellowish
4. brownish
5. olive
6. greenish
7. bluish
8. greyish

1. pink
2. red
3. yellow
4. brown
5. olive
6. green
7. blue
8. white
9. grey
0. black

Grain size
1. Very coarse (>60mm)
2. Coarse (2-60mm)
3. Medium (60μ-2mm)
4. Fine (2-60μ)
5. Very fine (<2μ)

Compressive strength
1. Very strong (>100MN/m²)
2. Strong (50-100MN/m²)
3. Mod. strong (12.5-50MN/m²)
4. Mod. weak (5-12.5MN/m²)
5. Weak (1.25-5MN/m²)
6. V. weak/hard (600-1250kN/m²)
7. Very stiff (300-600kN/m²)
8. Stiff (150-300kN/m²)
9. Firm (80-150kN/m²)
0. Soft (40-80kN/m²)

Method of determining compressive strength
1. Measured
2. Assessed

Rock type

Qualifying terms to describe rock

ROCK MASS INFORMATION

Fabric
1. Blocky
2. Tabular
3. Columnar

Block size
1. Very large (>8m³)
2. Large (0.2-8m³)
3. Medium (0.008-0.2m³)
4. Small (0.0002-0.008m³)
5. Very small (<0.0002m³)

State of weathering
1. Fresh
2. Slightly
3. Moderately
4. Highly
5. Completely
6. Residual soil

No. of major discontinuity sets

LINE SURVEYS TO DETERMINE DISCONTINUITY SPACINGS

	Plunge of line	Trend of line	Length of line (metres)	No. of fractures	Spacing	Remarks
Line 1						
Line 2						
Line 3						

Discontinuity spacing
1. Ext. wide (<2m)
2. Very wide (600mm-2m)
3. Wide (200-600mm)
4. Mod. wide (60-200mm)
5. Mod. narrow (20-60mm)
6. Narrow (6-20mm)
7. Very narrow (6mm)

Fig. 7.2 Rock mass description data sheet. (Anon, 1977; Copyright © 1977 by *Geological Society of London*)

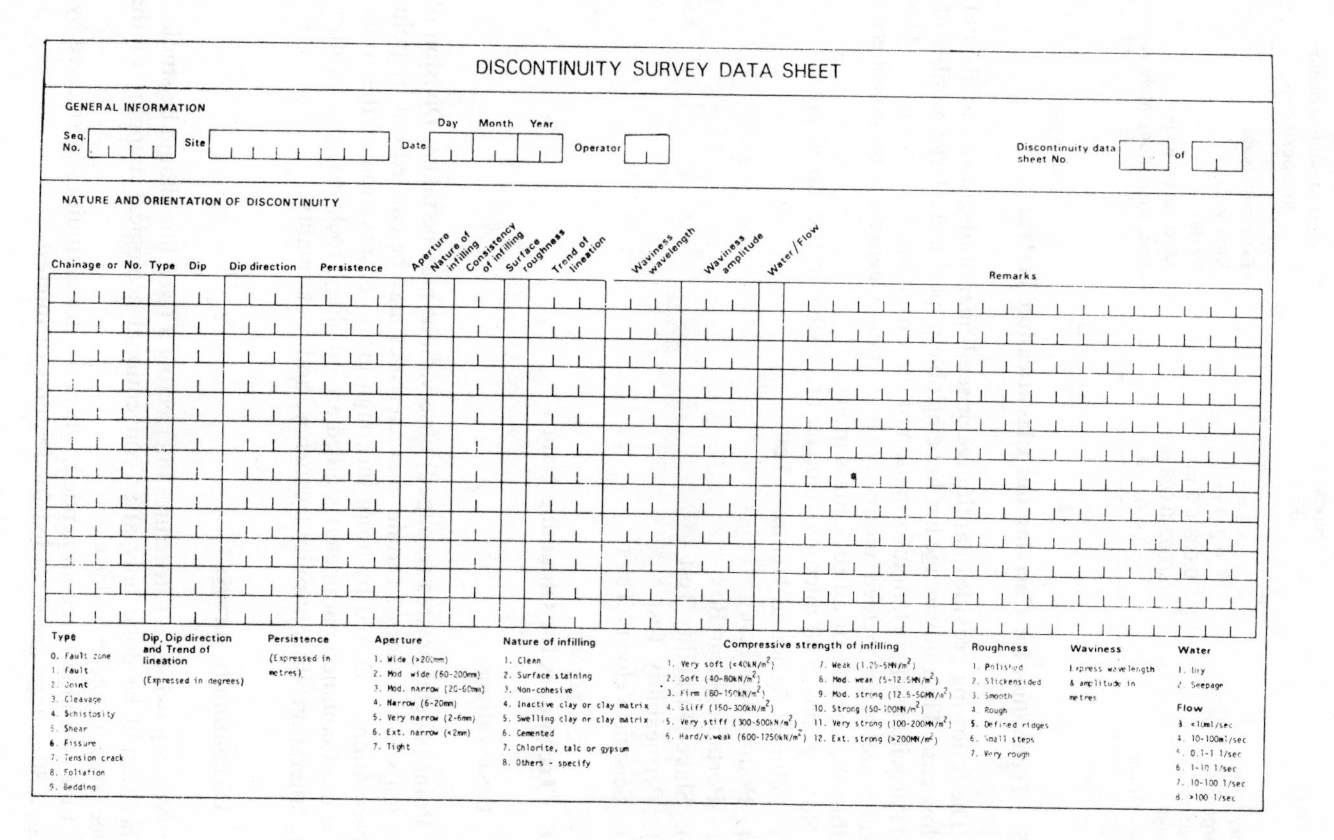

DISCONTINUITY SURVEY DATA SHEET

GENERAL INFORMATION

Seq. No. | Site | Date (Day, Month, Year) | Operator | Discontinuity data sheet No. ___ of ___

NATURE AND ORIENTATION OF DISCONTINUITY

Chainage or No.	Type	Dip	Dip direction	Persistence	Aperture	Nature of infilling	Consistency of infilling	Surface roughness	Trend of lineation	Waviness wavelength	Waviness amplitude	Water/Flow	Remarks

Type
0. Fault zone
1. Fault
2. Joint
3. Cleavage
4. Schistosity
5. Shear
6. Fissure
7. Tension crack
8. Foliation
9. Bedding

Dip, Dip direction and Trend of lineation
(Expressed in degrees)

Persistence
(Expressed in metres)

Aperture
1. Wide (>200mm)
2. Mod wide (60-200mm)
3. Mod. narrow (20-60mm)
4. Narrow (6-20mm)
5. Very narrow (2-6mm)
6. Ext. narrow (<2mm)
7. Tight

Nature of infilling
1. Clean
2. Surface staining
3. Non-cohesive
4. Inactive clay or clay matrix
5. Swelling clay or clay matrix
6. Cemented
7. Chlorite, talc or gypsum
8. Others - specify

Compressive strength of infilling
1. Very soft (<40kN/m^2)
2. Soft (40-80kN/m^2)
3. Firm (80-150kN/m^2)
4. Stiff (150-300kN/m^2)
5. Very stiff (300-500kN/m^2)
6. Hard/v.weak (600-1250kN/m^2)
7. Weak (1.25-5MN/m^2)
8. Mod. weak (5-12.5MN/m^2)
9. Mod. strong (12.5-50MN/m^2)
10. Strong (50-100MN/m^2)
11. Very strong (100-200MN/m^2)
12. Ext. strong (>200MN/m^2)

Roughness
1. Polished
2. Slickensided
3. Smooth
4. Rough
5. Defined ridges
6. Small steps
7. Very rough

Waviness
Express wavelength & amplitude in metres

Water
1. Dry
2. Seepage

Flow
3. <10ml/sec
4. 10-100ml/sec
5. 0.1-1 l/sec
6. 1-10 l/sec
7. 10-100 l/sec
8. >100 l/sec

Fig. 7.3 Discontinuity survey data sheet. (Anon, 1977; Copyright © 1977 by *Geological Society of London*)

Table 7.7 Block size.

Term	Block size	Equivalent discontinuity spacings in blocky rock
Very large	> 8 m^3	Extremely wide
Large	0.2–8 m^3	Very wide
Medium	0.008–0.2 m^3	Wide
Small	0.0002–0.008 m^3	Moderately wide
Very Small	< 0.0002 m^2	Less than moderately wide

7.5 Techniques for Carrying out Classification Tests

The following provide quantitative or semi-quantitative data rapidly and at low cost. Their advantage lies in being able to perform a large number of tests providing important information of the variability of a rock mass. The results are often yielded indirectly, and should generally be considered within the context of all rock mass indices.

Note: grains > 60 microns diameter are visible to the naked eye.

(a) Schmidt Rebound Hammer Test
(b) Point Load Test
(c) Portable Shear Box
(d) Slake Durability Test
(e) Permeability Test
(f) Seismic Velocity Tests.

7.6 Techniques for Obtaining Data

1 Observations

Point (single) or area (grouped) observations may be taken. Division of the mass into zones of certain characteristics may be carried out by point observation, followed by more thorough group observation of the zones. The most common parameters studied are lithology, rock strength, weathering/alteration, discontinuity frequency, and rock quality.

2 Discontinuity surveys

Where possible, all structures intersecting a fixed line should be studied. This should be done only after conventional geological mapping of the face/area has been undertaken.

Line sampling is recommended, where all discontinuities intersected by, e.g. a 30 m tape are recorded for features of interest.

In structurally complex areas, more subjective sampling methods may be undertaken the objects being, ultimately, to reveal the major structural trends at a site.

3 Recording and presentation of data

Data should be recorded on data sheets provided. The forms do not necessarily have to be fully completed; in some cases, collection of only a certain amount of information will prove adequate.

Geological data often show strong spatial interrelationships, and therefore lend themselves to cartographic and graphical representation. Angular relationships may be shown in rose diagrams, histograms and spherical projections. The last of these is strongly recommended for its more comprehensive portrayal of data, and its usefulness for statistical assessment.

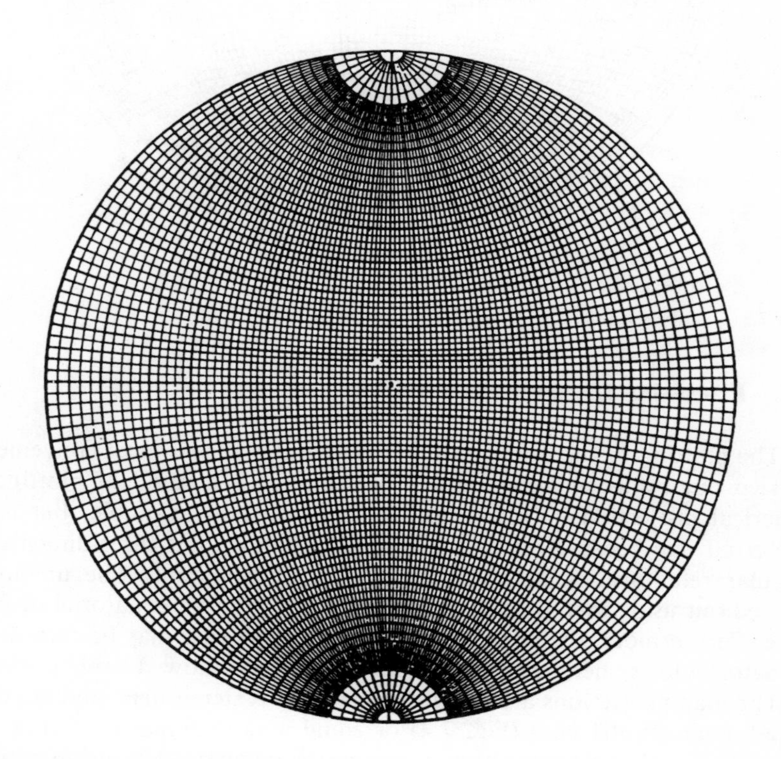

Fig. 7.4 Wulff Equal Angle equatorial stereonet.

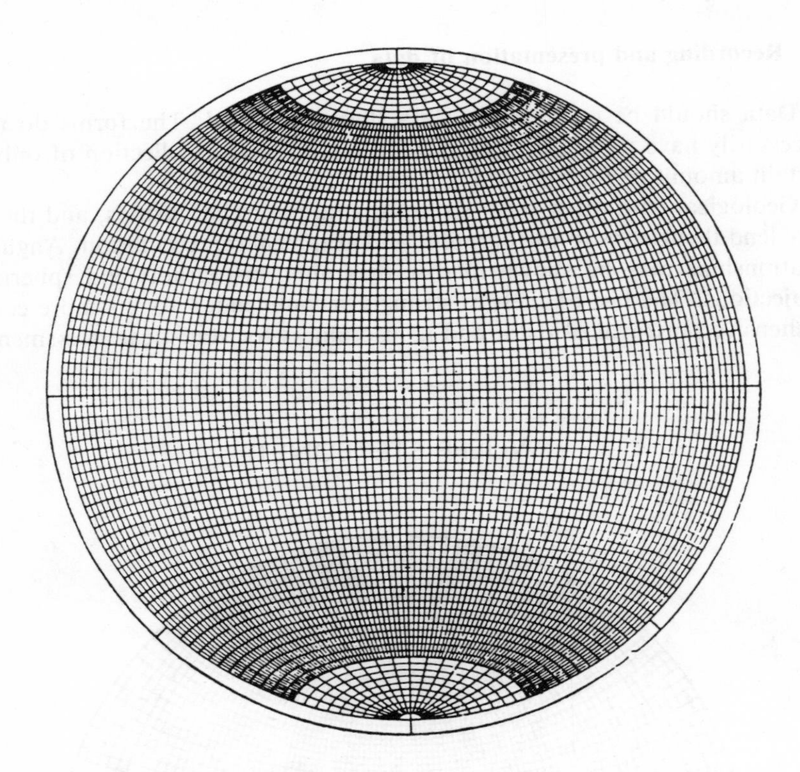

Fig. 7.5 Lambert Equal Area or Schmidt Equatorial Net.

7.7 Processing Discontinuity Data

The plotting of the orientation of structural planes is most conveniently carried out by using stereographic projections, a method of plotting on spherical projections. This is not the only method available, but is the preferred one because no part of the data is lost, and three-dimensional angular relationships between planes become apparent. The method is carried out using map projections, which may be of an equatorial or polar type. Furthermore, the upper or lower hemisphere case may be considered. Equatorial lower hemisphere conventions will be followed here.

The map projections are usually referred to as stereo-nets, and are of an equal angle (Wulff Net) (Fig. 7.4) or equal area (Schmidt Net) (Fig. 7.5) type. An equal angle projection preserves the angular relationship between planes plotted on the net, although some distortion of the graticule is necessary to achieve this. In an equal area net, the units of the graticule maintain equality of area throughout.

The procedure adopted is essentially determined by the amount of data which it is necessary to plot, and the use to which it is to be put. If angular relationships of a relatively small number of planes are to be studied, then cyclographic projections are plotted. These correspond to great circles on a map projection. The plane shown in Fig. 7.6 is to be plotted as a cyclographic projection. The plane is transferred to a sphere in which the equatorial projection with north at the top, has been "laid down" so that it appears to occupy the equatorial plane. It is clear from Fig. 7.7 what is meant by the lower hemisphere case, since only that part of the plane is

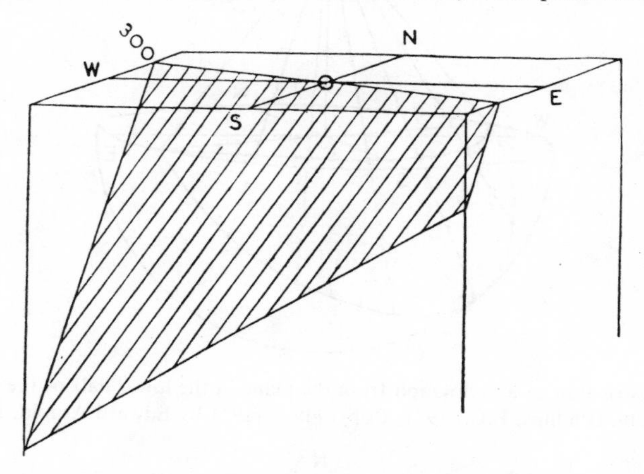

Fig. 7.6 An inclined plane displayed three dimensionally. (Phillips, F.C., 1971; Copyright © 1971 by Edward Arnold, London)

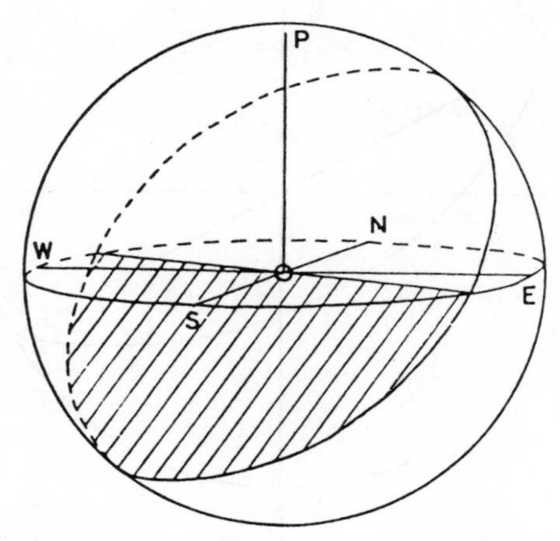

Fig. 7.7 The inclined plane shown in the reference sphere. (Phillips, F.C., 1971; Copyright © 1971 by Edward Arnold, London)

shaded. All the construction lines which enable the cyclographic projection of this plane to be constructed are projected from the lower hemisphere to point P as shown in Fig. 7.8. The trace which is produced on the equatorial plane is a great circle and represents the cyclograph of the plane shown in Fig. 7.6. Note that the south-westerly dip direction of the plane is marked by the direction of the maximum "bulge" of the cyclograph , and that the NW–SE strike is indicated by the "ends" of the cyclograph. The completed cyclograph of the plane, in the correct orientation is shown in Fig. 7.9.

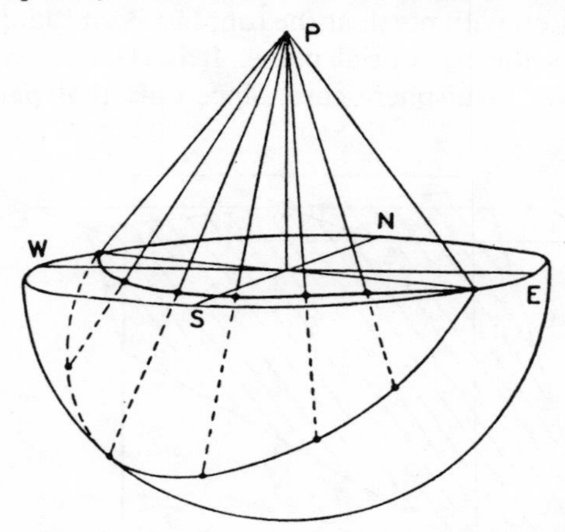

Fig. 7.8 Construction of a cyclograph from the plane in the lower half of the reference sphere. (Phillips, F.C., 1971; Copyright © 1971 by Edward Arnold, London)

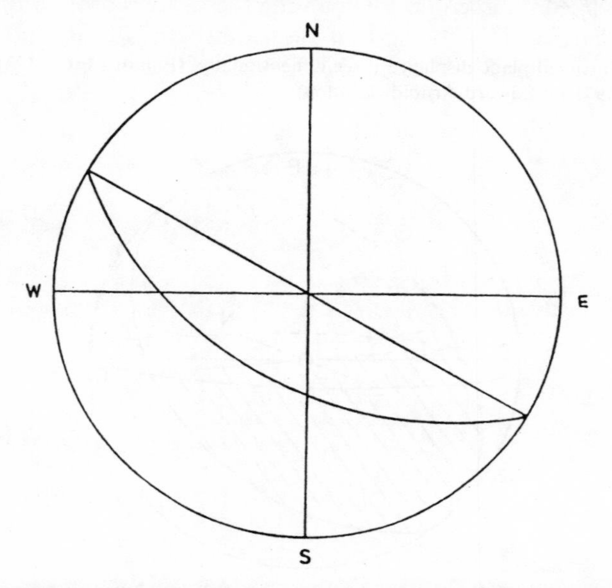

Fig. 7.9 The completed cyclograph shown as a stereographic projection. (Phillips, F.C., 1971; Copyright © 1971 by Edward Arnold, London)

The procedure adopted to arrive at this position to some extent depends on the form of the data to be plotted. As was explained in Chapter 3, geologists present structural data in one form, engineering geologists often in another:

Geologist:

Strike	Dip	Dip Direction
130°	20°	SW
032°	25°	SE
258°	48°	NNW

Engineering Geologist:

—	20°	220°
—	25°	122°
—	48°	348°

The readings of strike and dip direction are whole compass bearings from north.

To plot structural data, say a plane dipping at 40° in a direction of 130°, place a piece of tracing paper over the surface of a Wulff Net and push a pin through the centre of the net so that it allows the tracing paper to spin around it. Mark the position of north on the tracing paper. Mark the position of the dip direction, by counting the degrees around the primitive circle (circumference of the net), as shown in Fig. 7.10. Rotate the tracing paper until the point indicating the dip direction lies on the east or west point of the net, (Fig. 7.11). The dip is then marked off along the east-west diameter of the net. It is clear from Fig. 7.8 that the dip is measured from the primitive circle towards the centre, because the primitive circle itself acts

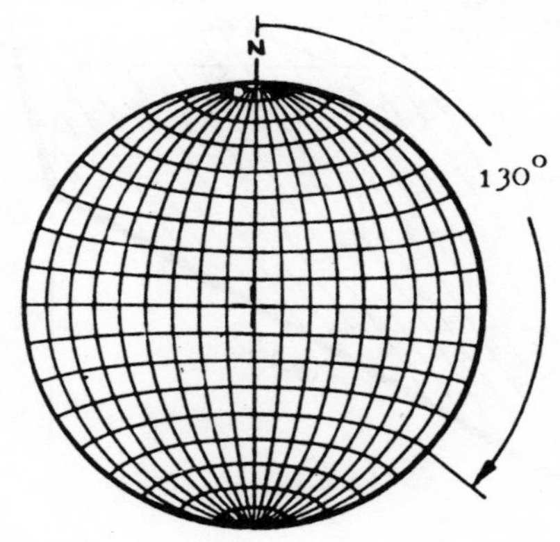

Fig. 7.10 First stage in the plotting procedure of cyclographic projections. (Hoek, E. and Bray, J.W., 1981; Copyright © 1981 by *Institution of Mining and Metallurgy*)

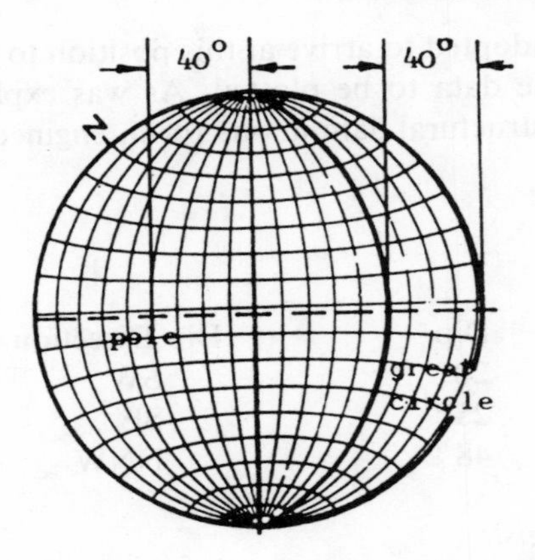

Fig. 7.11 Final stage in the plotting procedure of cyclographic projections. (Hoek, E. and Bray, J.W., 1981; Copyright © 1981 by *Institution of Mining and Metallurgy*)

as the cyclographic projection of a horizontal plane. While the trace is still in the rotated position, draw in the great circle from the north pole to the south pole on the net, through the dip mark on the east-west line, (Fig. 7.11). Rotate the trace back so that the position of north on the trace corresponds with north on the net, and the completed cyclographic projection is in its correct orientation.

When strike is given in the data, the procedure is slightly different. The strike is marked off as a whole compass bearing, in this case 040 °, and the mark is rotated to the north position. The dip is then marked off from the east point and the cyclograph drawn in. With either method, always check that the cyclograph "bulges" in the dip direction.

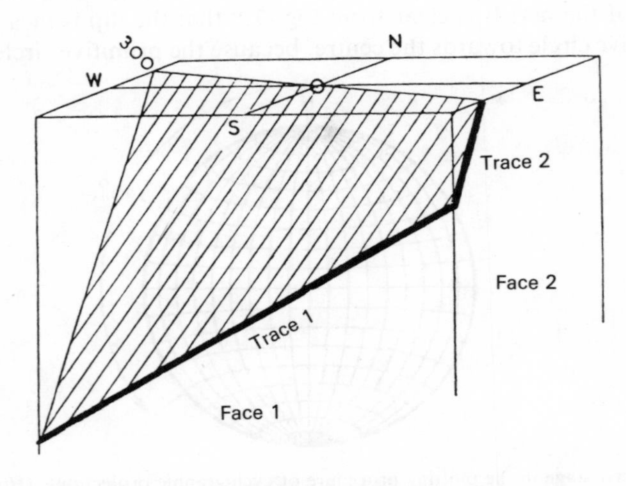

Fig. 7.12 Different attitudes of bedding traces on two faces.

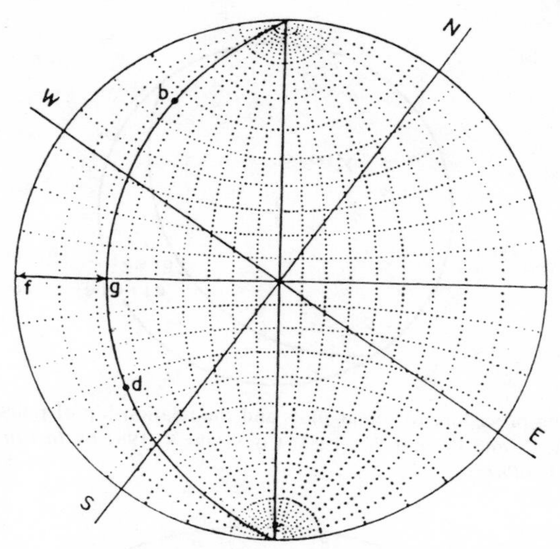

Fig. 7.13 Plotting of a cyclograph from apparent dip and dip direction data. (Phillips, F.C., 1971; Copyright © 1971 by Edward Arnold, London).

It is not always possible to measure the true dip and dip direction of a discontinuity and it becomes necessary to measure the apparent dip and apparent dip direction. This is simply the orientation of the two-dimensional trace of the discontinuity (Fig. 7.12). Where this is done, it is essential that at least two pairs of readings are taken. To plot these, the procedure is the same up to the drawing in of the final cyclograph, (Fig. 7.13). With the two points marked, rotate the trace until both points lie on the same great circle. This is then drawn in, the trace returned to north, and the cyclograph is that of the true dip and dip direction of the discontinuity.

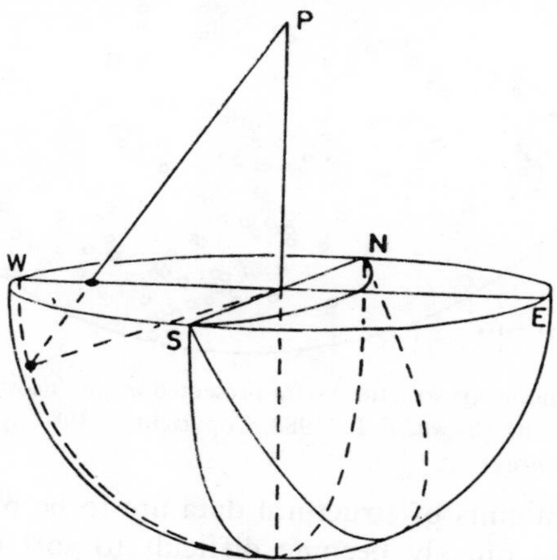

Fig. 7.14 Construction of the pole of the normal to a plane. (Phillips, F.C., 1971; Copyright © 1971 by Edward Arnold, London)

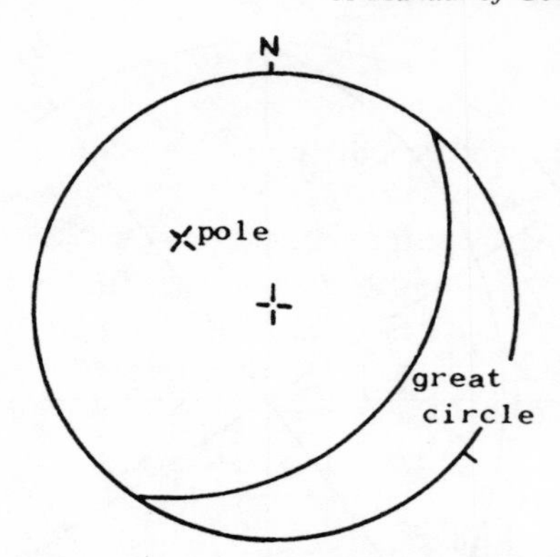

Fig. 7.15 Plotting procedure for polar plots and the relationship of poles to cyclographs. (Hoek, E. and Bray, J.W., 1981; Copyright © 1981 by *Institution of Mining and Metallurgy*)

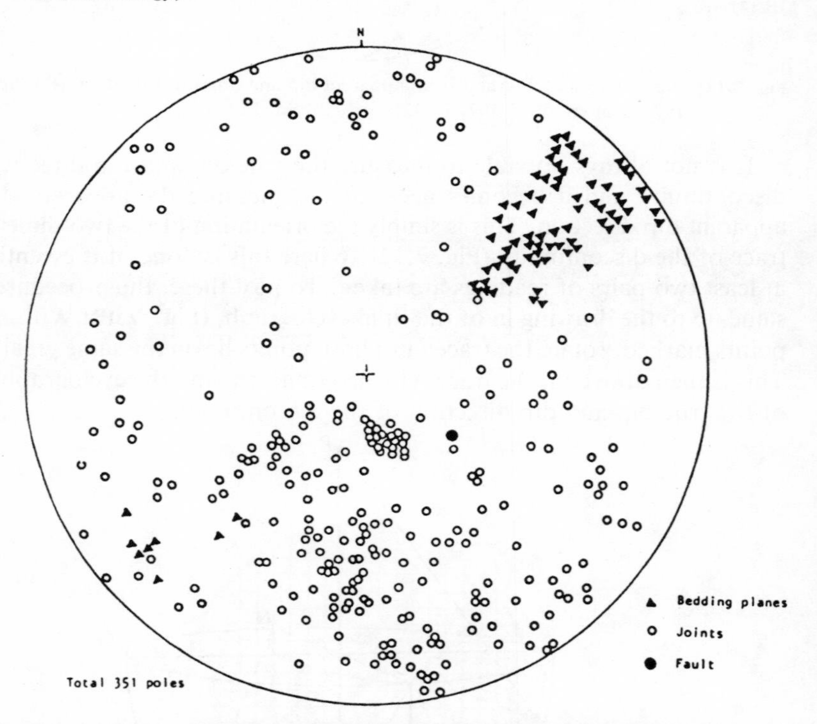

Fig. 7.16 A large amount of structural data presented as an unsorted polar projection. (Hoek, E. and Brown, E.T., 1980; Copyright © 1980 by *Institution of Mining and Metallurgy*)

When large amounts of structural data are to be plotted, cyclographic projections would quickly become difficult to sort out. It is therefore preferable to use an alternative method which will allow any structural grain within a rock mass to become apparent. This involves the plotting of the poles of the normals to planes, or polar plots (Fig. 7.14). These may be plotted by using the same technique as for cyclographic projections except that having followed the procedure up to the plotting of the dip, a different method is adopted. As is clear from Fig. 7.14, the pole is in the opposite half of the projections from that of its corresponding cyclographic projection (Fig. 7.15). Its position is 90° away from the centre of the cyclograph, measured with the great circle laying from north to south. Poles, as can be appreciated from Fig. 7.15, have the dip counted from the centre towards the primitive circle, so that the pole to a horizontal plane is at the centre of the net.

Points can be plotted in large quantities without becoming confusing (Fig. 7.16). If this is to be done, then a Schmidt equal area net should be used for the plotting, because sense can really only be made of the data when it is in a processed form. For this task, a counter is used (Fig. 7.17) the dimensions of which are closely related to the diameter (D) of the net. A grid with lines spaced at one-twentieth of the diameter of the net is placed beneath the tracing containing the poles. The poles are counted as shown in Fig. 7.18, that is, one of the circles on the counter is moved to each inter-

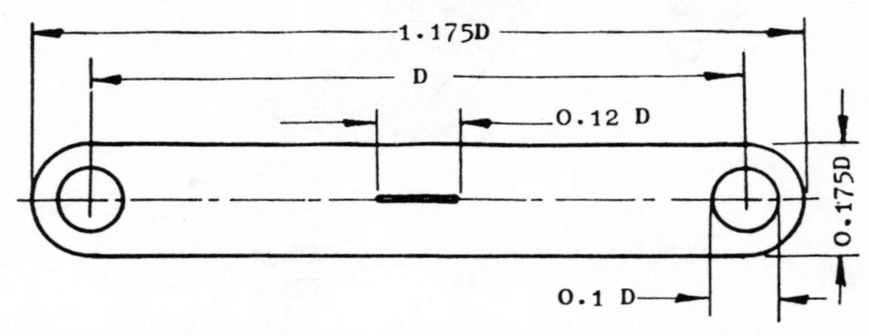

Fig. 7.17 The counter used for sorting polar data. (Hoek, E. and Bray, J.W., 1981; Copyright © 1981 by *Institution of Mining and Metallurgy*)

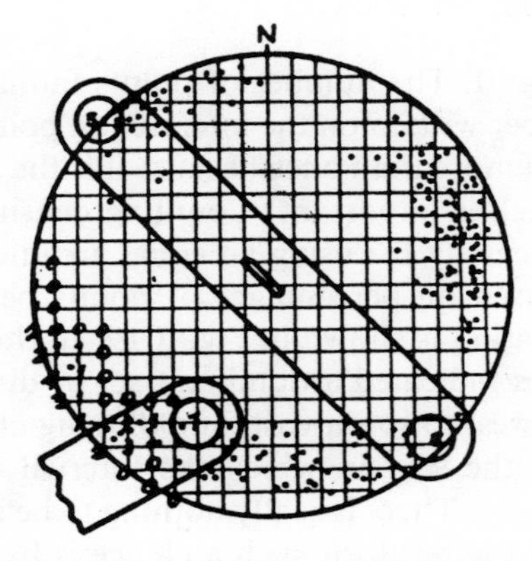

Fig. 7.18 The procedure for sorting polar data. (Hoek, E. and Bray, J.W., 1981; Copyright © 1981 by *Institution of Mining and and Metallurgy*)

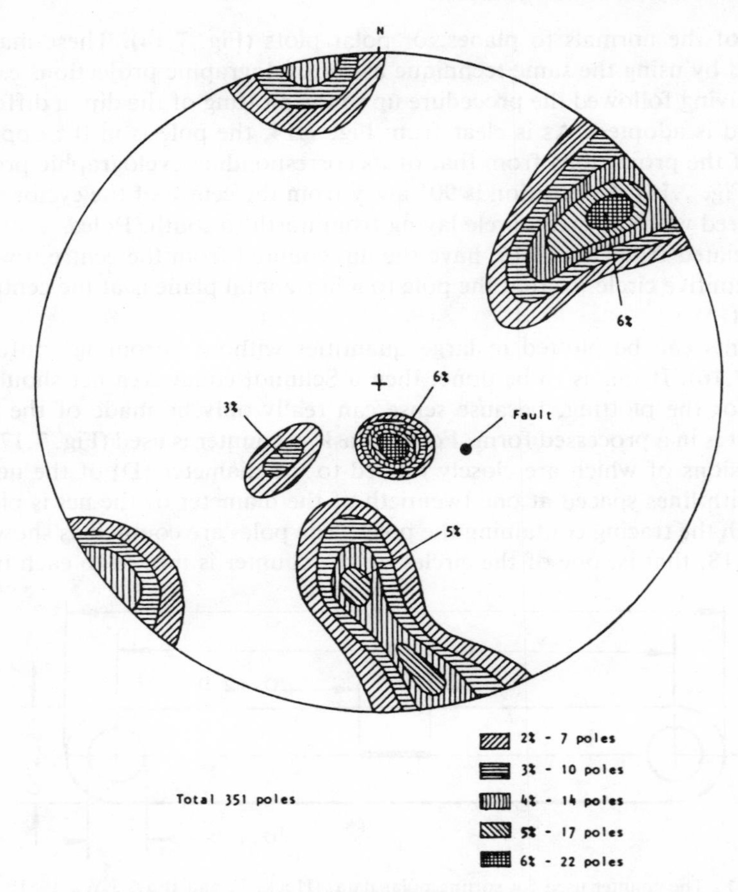

Fig. 7.19 A contoured polar plot emphasizing the directional anisotropy of the rock mass. (Hoek, E. and Brown, E.T., 1980; Copyright © 1980 by *Institution of Mining and Metallurgy*)

section point on the grid. The number of points falling within the circle is counted and the number written on the intersection point. It does not matter that poles are counted more than once because it is the density of poles, and not their number which is important. Counting densities of poles per unit area highlights the necessity for using an equal area net. The plot is subsequently contoured based on percentages of points per area. A final contoured equal area diagram is shown in Fig. 7.19. In this example, the main concentration of poles indicated a dominant set of discontinuities striking approximately north-west to south-east, and dipping at a moderate angle to the south-west. Note the way in which the interval of contours of pole density have been chosen. There is really nothing to be gained by having too close an interval, and the result of such a choice is likely to make the plot confusing.

Experience indicates that stereographic projections are an excellent way to present structural data, especially when analysing stability of slopes (Hoek and Bray, 1981), tunnels and underground chambers cut in rock (Hoek and Brown, 1980). Methods of stability analysis based on stereographic projection techniques are widely used in geotechnical engineering. The technique is far more flexible, especially when used to display three-dimensional situations, than conventional engineering drawing methods. The two methods are quite different, however, but the unfamiliarity with the principles and form of stereographic projections should not be an excuse to ignore the value of this method. Further information on the background to stereographic projections can be found in Phillips (1971).

REFERENCES

Andric, M., Roberts, G. T., and Tarvydas, R. K., 1976. Engineering Geology of the Gordon Dam, West Tasmania. *Quarterly Journal of Engineering Geology*, Vol. 9, No. 1, pp. 1–24.

Anon, 1970. The Logging of Rock Cores for Engineering Purposes. *Quarterly Journal of Engineering Geology*, Vol. 3, No. 1, pp. 1–24.

Anon, 1972. The Preparation of Maps and Plans in Terms of Engineering Geology. *Quarterly Journal of Engineering Geology*, Vol. 5, No. 4, pp. 292–382.

Anon, 1977. The Description of Rock Masses for Engineering Purposes. *Quarterly Journal of Engineering Geology*, Vol. 10, No. 4, pp. 355–388.

Barton, N., 1974. A Review of the Shear Strength of Filled Discontinuities. Norwegian Geotechnical Institute Publication No. 105, pp. 1–38.

Bell, F. G., 1981. Engineering Properties of Soils and Rocks. Butterworths, London, 149 pp.

Broch, E. and Franklin, J. A., 1972. The Point-Load Strength Test. *International Journal of Rock Mechanics and Mining Science*, Vol. 9, pp. 669–697.

Brunsden, D., 1973. The Application of Systems Theory to the Study of Mass Movement. *Estratto da Geologia Applicata e Idrogeologia, Bari*, Vol. 8, Part 1, pp. 185–207.

Burnett, A. B. and Fookes, P. G., 1974. A Regional Engineering Geological Study of the London Clay in the London and Hampshire Basins. *Quarterly Journal of Engineering Geology*, Vol. 7, No. 3, pp. 257–295.

Chorley, R. J., 1962. Geomorphology and General Systems Theory. United States Geological Survey Professional Paper, 500-B, 10 pp.

Clark, M. W., 1979. Marine Processes. In: Process in Geomorphology, ed. C. Embleton and J. Thornes. Edward Arnold, London, 436 pp.

Dearman, W. R., Money, M. S., Strachan, A. D., Coffey, J. R. and Marsden, A., 1979. A Regional Engineering Geological Map of the Tyne and Wear County. N.E. England. *Bulletin of the International Association of Engineering Geology*, No. 19, pp. 5–17.

Fookes, P. G., Dearman, W. R. and Franklin, J. A., 1971. Some Engineering Aspects of Rock Weathering with Field Examples from Dartmoor and Elsewhere. *Quarterly Journal of Engineering Geology*, Vol. 4, No. 3, pp. 139–185.

Franklin, J. A., Broch E., and Walton, G., 1971. Logging the Mechanical Character of Rock. Transactions, Section A, of the Institution of Mining and Metallurgy, Vol. 80, pp. A1–A9.

Franklin, J. A. and Chandra, R., 1972. The Slake-Durability Test. *International Journal of Rock Mechanics and Mining Science*, Vol. 9, pp. 325–341.

Glavin, C. J., 1968. Breaker Type Classification on Three Laboratory Beaches. *Journal of Geophysical Research*, Vol. 73, pp. 3651–3659.

Griggs, D. T., 1936. The Factor of Fatigue in Rock Exfoliation. *Journal of Geology*, Vol. 44, pp. 781–796.

Gupta A., Rahman A., and Wong, P. P., 1980. Cenozoic Alluvium in Singapore: Its Palaeo-Environment Reconstructed from a Section. *Singapore Journal of Tropical Geography*, Vol. 1, No. 2, pp. 40–46.

Harvey, P., 1982. Estuary Dams and Dykes for Storage of Fresh Water in Marine Environment. *Regional Symposium on Underground Works and Special Foundations*, Singapore, pp. 1–6.

Hoek, E. and Bray, J. W., 1981. Rock Slope Engineering, 3rd edition. Institution of Mining and Metallurgy, London, 309 pp.

Hoek, E. and Brown, E. T., 1980. Underground Excavations in Rock. Institution of Mining and Metallurgy, London, 527 pp.

Leopold, L. B., Wolman, M. G. and Miller, J. P., 1964. Fluvial Processes in Geomorphology. W. H. Freeman and Company, San Francisco, 522 pp.

Linton, D. L., 1955. The Problem of Tors. *Geographical Journal*, Vol. 121, pp. 407–487.

Lundegard, P. D. and Samuels, N. D., 1980. Field Classification of Fine-Grained Sedimentary Rocks. *Journal of Sedimentary Petrology*, Vol. 50, No. 3, pp. 781–786.

Moh, Z-C. and Mazhar, M. F., 1969. Effects of Method of Preparation in Index Properties of Lateritic Soils. Proceedings, speciality session on Engineering Properties of Lateritic Soils, Vol. 1. *7th International Conference on Soil Mechanics and Foundation Engineering*, Mexico City.

Muller, L., 1964. Rock Slide in the Vaiont Valley. *Rock Mechanics and Engineering Geology*, Vol. 2, Nos. 3–4, pp. 148–212.

Muller, L. 1968. New Considerations on the Vaiont Slide. *Rock Mechanics and Engineering Geology*, Vol. 6, Nos. 1–2, pp. 1–91.

Nixon, I. K. and Skipp, B. O., 1957a. Airfield Construction in Overseas Soils. Part 6, Tropical Red Clays. *Proceedings, Institution of Civil Engineers*, Vol. 36, pp. 275–292.

Nixon, I. K. and Skipp, B. O., 1957b. Airfield Construction in Overseas Soils. Part 5, Laterite. *Proceedings, Institution of Civil Engineers*, Vol. 36, pp. 253–275.

Nossin, J. J. and Levelt, T. W. M., 1967. Igneous Rock Weathering on Singapore Island. *Zeitschrift für Geomorphologie*, Vol. 11, No. 1, pp. 14–35.

Ola, S. A., 1978. Geotechnical Properties and Behaviour of Stabilised Lateritic Soils. *Quarterly Journal of Engineering Geology*, Vol. 11, No. 2, pp. 145–160.

Ollier, C. D., 1969. Weathering. Longman, London. Reprinted 1975, 304 pp.

The Open University, 1972. Field Relations. 5-23, Block 2 Geology, The Open University Press, Bletchley, 70 pp.

Oxburgh, R. E., 1974. The Plain Man's Guide to Plate Tectonics. *Proceedings of the Geologists' Association*, Vol. 85, No. 3, pp. 299–357.

Patton, F. D., 1966. Multiple Modes of Shear Failure in Rock. Proceedings, First International Congress of the International Society of Rock Mechanics, Lisbon, Vol. 1, pp. 509–513.

Peltier, L. C., 1950. The Geographical Cycle in Periglacial Regions as it is Related to Climatic Geomorphology. *Annals of the Association of American Geographers*, Vol. 40, pp. 214–236.

Phillips, F. C., 1971. The Use of Stereographic Projection in Structural Geology, 3rd edition. Edward Arnold, London, 90 pp.

Pitts, J., 1979. Morphological Mapping in the Axmouth-Lyme Regis Undercliffs, Devon. *Quarterly Journal of Engineering Geology*, Vol. 12, No. 3, pp. 205–217.

Pitts, J., 1983a. Geomorphological Observations as Aids to the Design of Coast Protection Works on a Part of the Dee Estuary. *Quarterly Journal of Engineering Geology*, Vol. 16, No. 4, pp. 291–300.

Pitts, J., 1983b. The Form and Causes of Slope Failures in an Area of West Singapore Island. *Singapore Journal of Tropical Geography*, Vol. 4, No. 2, pp. 162–168.

Pitts, J., 1983c. Faults and other Shears in Bedded Pleistocene Deposits on the Wirral, United Kingdom. *Boreas*, Vol. 2, pp. 138–144.

Pitts, J., 1983d. The Nature and Origin of Bubble Structures in Borehole Samples of Glacial Sands from Caldy, Wirral. *Boreas*, Vol. 12, No. 4, pp. 247–251.

Pitts, J., 1983e. The Origin, Nature and Extent of Recent Deposits in Singapore. Proceedings, International Seminar on Construction Problems in Soft Soils, Singapore, pp. JP-1 to JP-18.

Pitts, J., 1984a. A Review of Geology and Engineering Geology in Singapore. *Quarterly Journal of Engineering Geology*, Vol. 17, No. 2, pp. 93–101.

Pitts, J., 1984a. A Survey of Engineering Geology in Singapore. *Geotechnical Engineering*, Vol. 15, No. 1, pp. 1–20.

Powell, C. McA., 1979. A Morphological Classification of Rock Cleavage. In: T. H. Bell and R. H. Vernon (editors), Microstructural Processes during Deformation and Metamorphism, *Tectonophysics*, Vol. 58, No. 1, pp. 21–34.

Price, D. G., Malkin, A. B. and Knill, J. L. 1969. Foundations of Multistorey Blocks on the Coal Measures with Special Reference to Old Mine Workings. *Quarterly Journal of Engineering Geology*, Vol. 1, No. 4, pp. 271–322.

Rahman, A. and Gupta, A., 1980. Rainfall Runoff Relation. Regional Training Course on Stormwater Management, University of Singapore and UNESCO, pp. 37–72 (mimeographed).

Sherrell, F. W., 1970. Some Aspects of the Triaseic Aquifer in East Devon and West Somerset. *Quarterly Journal of Engineering Geology*, Vol. 2, No. 4, pp. 255–286.

Sinclair, T. J. E., 1980. Strength and Compressibility Characteristics of a Lateritic Residual Soil. Proceedings, 6th Southeast Asian Conference on Soil Engineering, Taipei, pp. 113–125.

Skempton, A. W., 1964. The Long-term Stability of Clay Slopes. *Geotechnique*, Vol. 14, No. 2, pp. 77–102.

Skempton, A. W., 1970. First Time Slides in Overconsolidated Clays. *Geotechnique*, Vol. 20, No. 4, pp. 320–324.

Strakhov, N. M., 1967. Principles of Lithogenesis, Vol. 1. Consultants Bureau, N.Y., Oliver and Boyd, London, Trans. J.P. Fitzsimmons, ed. S. I. Tomkieff and J. E. Hemingway, 245 pp.

Terzaghi, K., 1962. Stability of Steep Slopes on Hard Unweathered Rock. *Geotechniquė*, Vol. 12, No. 2, pp. 251–270.

Varnes, D. J., 1978. Slope Movement Types and Processes. In: Landslides, Analysis and Control (ed. R. L. Schuster and R. J. Krizek). Special Report 176, Transportation Research Board, National Research Council, Washington D.C., pp. 11–33.

Walton, E. K., 1971. Looking Back Through Time. Chap. 13 in "Understanding the Earth", edited by I. G. Glass, P. J. Smith and R. C. L. Wilson, The Artemis Press, Sussex.

INDEX